MW00835389

Linear Quadratic Control: An Introduction

PETER DORATO, CHAOUKI T. ABDALLAH.
Department of Electrical and Computer Engineering
University of New Mexico.

VITO CERONE.
Dipartimento di Automatica e Informatica
Politecnico di Torino.

KRIEGER PUBLISHING COMPANY
MALABAR, FLORIDA
2000

Original Edition 1995
Reprint Edition 2000 w/corrections

Printed and Published by
KRIEGER PUBLISHING COMPANY
KRIEGER DRIVE
MALABAR, FLORIDA 32950

Library of Congress Cataloging-in-Publication Data

Dorato, Peter.
Linear quadratic control : an introduction / Peter Dorato, Chaouki T. Abdallah, Vito Cerone.
p. cm.
Originally published: Prentice-Hall. 1995.
Includes bibliographical references and index.
ISBN 1-57524-156-0 (hc : alk. paper)
1. Linear control systems. 2. Control theory. I. Abdallah, C. T. (Chaouki T.) II. Cerone, Vito. III. Title.

TJ220 .D66 2000
629.8'32—dc21

00-030433

10 9 8 7 6 5 4 3 2

Alla mia famiglia
P.D.

To My three C's
C.T.A.

A Laura
V.C.

Preface

Linear quadratic control theory has become an important tool for feedback system design for a number of reasons:

- It provides an analytical approach to the design of multivariable feedback systems.
- It applies to both time-invariant and time-varying systems.
- It can deal directly with disturbance signals and sensor noise.
- It can deal with finite and infinite time intervals.
- It has ample software for practical design problems.

The purpose of this book is to provide an introduction to linear quadratic theory beyond the typical chapter on the subject found in texts on optimal or multivariable control. Of course, as an introductory text, the material here is not as detailed as that found in advanced books devoted exclusively to the subject. Basically, the goal is to fill the gap between these two extremes. Most of the material covered is fairly standard, but we have attempted to place more emphasis on the robustness issue and on stochastic control than found in other introductions to the subject.

At the University of New Mexico, linear quadratic control theory takes up about 50 percent of a one-semester graduate course on multivariable control required of all graduate students who wish to specialize in control systems. Most of the material in the first six chapters is generally included in the course, and based on time availability, parts of the remaining chapters are also included. Since it is rare to find graduate programs whose courses on the subject are much more extensive than this, this text should fit a large number of graduate programs in the systems area. It should also be of value to practicing engineers who need a relatively simple, but self-contained, introduction to the subject. The sections on MATLAB® were written especially for the practicing engineer, who must have available numerical algorithms for practical multivariable system design.

The first author would like to acknowledge the support received from the National Science Foundation, under grant INT-9016501, for his sabbatical leave for the academic year 1991–92, during which time the first draft of this book was written. He would also like to thank the Centro di Elaborazione Numerale dei Segnali, Consiglio Nationale delle Ricerche, for their invitation to visit the Politecnico di Torino during this period of time. All authors would like to acknowledge support from their respective departments, the Department of Electrical and Computer Engineering at the University of New Mexico and the Dipartimento di Automatica e Informatica at the Politecnico di Torino.

The authors would like to express their gratitude for the Foreword written by Professor Michael Athans of the Massachusetts Institute of Technology. Professor Athans is one of the leading contributors to the theory and applications of linear quadratic control, and his comments on the past and future of the field are greatly appreciated. The authors would also like to thank Professor Giovanni Fiorio, at the Politecnico di Torino, for his classroom testing of the first draft and his valuable input on the draft contents, and the students at the University of New Mexico, where the book was tested several years before publication. Finally, the authors would like to express their appreciation to the following colleagues who reviewed the original manuscript and made many useful suggestions for improving it: Dr. R. Chiang, Jet Propulsion Laboratories; Professor C. Hollot, University of Massachusetts; Dr. S. M. Joshi, NASA Langley Center; Professor F. Lewis, University of Texas at Arlington; Professor D. Petersen, University of New Mexico; and Professor B. Shafai, Northeastern University. The help of Professor D. Bader of the University of New Mexico is also greatly appreciated.

P. D, C.T.A, and V. C

Contents

Foreword

The design of optimal feedback control systems for linear plants by using quadratic penalties on the state and control variables represents one of the most studied classes of problems in dynamic deterministic and stochastic optimal control theory. Results are available for both the time-varying and the time-invariant cases, and for both the continuous-time and discrete-time system models. From a theoretical perspective, the so-called Linear-Quadratic-Regulator (LQR) problem and the Linear-Quadratic-Gaussian (LQG) problem (the H^2 problem, as it is called these days) offer elegant methodologies amenable to numerical solution using general-purpose CAD software. Deep understanding of these problems is essential for more advanced studies related to their frequency-domain properties, robustness, and worst-case performance guarantees (the so-called H^∞ design methodology).

This book represents a welcome addition to the introductory texts we have available on these topics. It discusses classical LQR and LQG ideas and provides bridges to their more recent offspring, such as Linear-Quadratic-Gaussian/Loop-Transfer-Recovery (LQG/LTR) and introductory H^∞ synthesis. The analytical material is interleaved with MATLAB® software discussion so that nontrivial problems can be numerically solved and simulated to illustrate the design methodologies. It is easy to read; as an introductory text, it avoids extensive mathematical proofs.

Linear-quadratic design concepts have been around since the late fifties. LQR and LQG methods represent the first attempts at systematic design of multivariable control systems using both deterministic and stochastic dynamic optimal control ideas. Indeed, several early versions of CAD software appeared in the early sixties (in retrospect, many of them generating numerical solutions of dubious quality), signifying the confidence of the field that the stuff was indeed useful for practical designs. The basic concepts were then extended into dynamic differential games (providing a time-domain based beginning of H^∞ optimal control), decentralized control strategies (using regular and singular perturbations ideas), and design of fixed-structure compensators. Later on, the availability of cheap and reliable microprocessors; the emergence of multivariable frequency-domain insights and stability-robustness tests; the reformulation of the optimization problem in the frequency domain, including the use of frequency-dependent weights; and the development of high-quality design software packages jointly contributed to the practical utilization of the theory. Nevertheless, it is incredible that new concepts, methodologies, and insights are still being generated after forty years of research on the subject. This vitality can be explained only by the constant interplay between theory and applications, as well as the solid analytical foundation of the field. Unlike many recent "design fads" (which promise to control everything without bothering with mathematics, plant models, precise constraints and specifications, and stability and robustness guarantees), the wise use of the powerful analytical and software control-theoretic design methodologies, has had a most significant impact in numerous application areas. As we now tackle the tough design questions in the areas of multivariable robust control, nonlinear control, adaptive control, and decentralized systems, it is comforting to have a wealth of ideas originating with linear quadratic control to guide our thinking and to provide a solid foundation for the advances still to be made. I firmly believe that control theory remains an exciting and vibrant field of research, with numerous application areas, and that significant discoveries await the persistent and inquiring researcher.

Michael Athans
Massachusetts Institute of Technology
Cambridge, Massachusetts
September 1993

Notation

- A^{-1}: Inverse of square matrix A
- A': Transpose of matrix A
- $P_2 \geq P_1$: $x'(P_2 - P_1)x \geq 0$, for all vectors x, where $P_1' = P_1$ and $P_2' = P_2$
- $P_2 > P_1$: $x'(P_2 - P_1)x > 0$, for all vectors $x \neq 0$, where $P_1' = P_1$ and $P_2' = P_2$
- $[t, T]$: Closed interval $t \leq \tau \leq T$
- $I_{n \times n}$: $n \times n$ unit matrix
- $0_{n \times n}$: $n \times n$ zero matrix
- $E\{x\}$: Expected value of the random variable x
- $\lambda(A)$: Eigenvalue of square matrix A
- $\bar{\sigma}(A) = \sqrt{\lambda_{max}(A^*A)}$: Largest singular value of A
- $\underline{\sigma}(A) = \sqrt{\lambda_{min}(A^*A)}$: Smallest singular value of A
- $\rho(A)$: Spectral radius of square matrix A (given by the largest magnitude of its eigenvalues)
- $\text{diag}(\lambda_1, \cdots, \lambda_n)$: Diagonal matrix with elements $\lambda_1, \cdots, \lambda_n$ along the diagonal
- $det(A)$: Determinant of square matrix A
- W^*: Complex conjugate of the transpose of matrix W
- $tr(A)$: Trace of square matrix A (sum of diagonal terms)

- $\| A \|$: Norm of matrix A. Unless otherwise stated, the norm is taken to be the induced Euclidean norm, i.e. $\| A \| = \sqrt{\lambda_{max}(A^* A)}$
- $\| F(s) \|_\infty$: H^∞ norm of the matrix function $F(s)$
- $\| F(s) \|_2$: H^2 norm of the matrix function $F(s)$
- $\frac{\partial V(x)}{\partial x}$: Gradient of the scalar function $V(x)$ with respect to the vector x, taken as a column vector $\left[\frac{\partial V(x)}{\partial x_1} \;\; \frac{\partial V(x)}{\partial x_2} \;\; \cdots \;\; \frac{\partial V(x)}{\partial x_n} \right]'$

Chapter 1

Introduction

In this chapter we first give a brief historical sketch of the optimal linear quadratic problem. We then define the prerequisites required for the material presented. Finally, we discuss the organization and intent of the book.

1.1 Historical Sketch

The linear quadratic control problem has its genesis in the work of Wiener on mean-square filtering for weapons fire control during World War II. His results were published shortly after the war in [203] and [204]. *The term "linear" comes from the fact that the systems considered were assumed linear, and the term "quadratic" comes from the use of performance measures that involve the square of an error signal.* However, the term **"linear quadratic"** did not appear in the literature until the late fifties. Prior to that time the problem was commonly referred to as the **mean-square** control problem.

Wiener solved the problem of designing filters that minimize a mean-square-error criterion (performance measure) of the form

$$V = E\{e^2(t)\} \tag{1.1}$$

where $E\{X\}$ denotes the expected value of the random variable X. For a linear time-invariant system with transfer function $T(s)$, a mean-square performance measure of the type given in (1.1) may be evaluated in the steady state by the integral formula

$$E\{e^2(t)\} = \frac{1}{2\pi}\int_{-\infty}^{\infty} |T(j\omega)|^2 \, \Phi(\omega) d\omega \tag{1.2}$$

where $\Phi(\omega)$ represents the input-noise spectrum. Wiener used spectral factorization to minimize such integrals. This analytical approach to filtering was applied to the design of feedback systems by Newton, Gould, and Kaiser in their now-classic text of 1957 [151]. This analytical approach had a great deal of theoretical appeal, since it avoided the trial-and-error methods associated with the Nyquist stability criterion and loop shaping. It also allowed the designer to take into account such practical issues as sensor noise and control input saturation, at least as measured in the mean-square sense.

From Parseval's theorem, we also know that if $E(s)$ is the transform of $e(t)$, then

$$\int_0^\infty e^2(t)\, dt = \frac{1}{2\pi}\int_{-\infty}^{\infty} |E(j\omega)|^2\, d\omega \tag{1.3}$$

Finally, the integral-square term in the time domain in (1.3) can be generalized to an integral-quadratic term of the form

$$\int_0^\infty e'(t)Qe(t)\, dt \tag{1.4}$$

In 1958, Kalman and Koepcke [104] used this type of generalized performance measure and state-variable models to synthesize a state-feedback controller for a sampled-data system, using what we now call **linear-quadratic control theory** for the first time. The sampled-data results were extended to continuous-time results shortly thereafter by Kalman [100] in the United States and Letov [119] in the USSR. In particular, the problem considered in [100] was that of minimizing the **quadratic performance measure**

$$V = \int_t^T (x'Qx + u'Ru)\, d\tau + x'(T)Mx(T) \tag{1.5}$$

subject to the **linear dynamical constraint**

$$\dot{x} = Ax + Bu \tag{1.6}$$

which is now known as the optimal **linear-quadratic-regulator** problem, or LQR problem. This problem was reduced to the solution of a **matrix Riccati differential equation**, and the period 1960–70 witnessed many theoretical studies devoted to the solution of this equation and the LQR problem. The new LQR theory offered a number of advantages over existing design techniques. In particular,

- It allowed for optimization over finite time intervals (the mean-square frequency-domain approach was limited to infinite optimization intervals).
- It was applicable to time-varying systems (the frequency-domain approach was limited to time-invariant systems).
- It dealt in a relatively simple way with multivariable systems (the mean-squared approach is extendable to multivariable systems, but only with some difficulty). Actually, it was not until relatively recently (1976) that the mean-squared theory was extended in a rigorous way to multivariable systems (see Youla, Bongiorno, and Jabr [213]).

However, the LQR theory did not deal with two central issues associated with the design of feedback-control systems: plant uncertainty and sensor noise. In addition, the LQR theory required knowledge of the state of the system which is often unavailable. In 1961, Kalman and Bucy [103] introduced a state-variable version of the Wiener filter. This version allowed for the optimal estimation of the state of a system from noisy measurements of the system's output. The optimal estimation problem was also reduced to the solution of a Riccati equation. In 1963, Gunckel and Franklin [84] used the Kalman–Bucy filter to design state-estimate feedback systems that minimized the expected value of a quadratic performance measure for sampled-data systems. *The linear-quadratic problem with noisy-output observations was reduced to the solution of two decoupled Riccati equations, by means of a separation theorem that basically states that the problem of optimal control with noisy-output measurements can be separated into the problem of optimal control with state feedback and the problem of optimal state estimation.* Since this stochastic linear-quadratic problem was based on the assumption that all the stochastic disturbance signals were Gaussian, it became popularly known as the **linear-quadratic-Gaussian (LQG)** problem. A rigorous proof of the separation theorem for continuous-time systems was given by Wonham [210] in 1968. By the early seventies, the LQG problem had reached a high level of development, and in 1971, a special issue of the *IEEE Transactions on Automatic Control* [11], edited by Athans, was published on the subject.

Of course, the issue of plant uncertainty could also be dealt with using Gaussian stochastic models (stochastic differential equations), and a fairly extensive theory of stochastic control theory was being developed at about this same time. Indeed, in 1967 Kushner published a book on the subject [112], summarizing most of the theory available at the time. Basically, it was possible to show that LQR problems where plant uncertainty is modeled by appropriate stochastic disturbance signals can be reduced to the solution of a "modified" Riccati equation [209]. On the other hand, the problem of plant uncertainty modeled by deterministic plant perturbations was more complex. In 1964, Kalman [101] was able to show for single-input/single-output systems that the Nyquist plot of the loop gain for an optimal LQR feedback system never entered a circle of radius one and centered about the minus one (-1) point. This implied some very strong gain and phase margins for the optimal closed-loop system, i.e., an infinite increasing gain margin and a phase margin of plus or minus 60 degrees. *However, in 1978, Doyle [56] was able to demonstrate that optimal LQG output feedback systems could have arbitrarily small stability margins.* A year later, Doyle and Stein [60] proposed a procedure that could asymptotically recover for output feedback, the desirable loop-transmission properties of the optimal LQR state-feedback solution. Their approach was based on introducing additional "artificial" noise in the system's dynamics and letting the noise intensity become infinite. They also required that the plant be minimum phase. The concept of design based on recovering certain target loop-transmission properties is now commonly referred to as **loop-transfer-recovery (LTR) design**, and the recovery approach proposed by Doyle and Stein is referred to as **LQG/LTR design**. The LQG/LTR approach requires a compromise between loop recovery (robust-stability margins) and performance-measure degradation.

In 1972, Chang and Peng [41] proposed a direct state-space approach to the problem of plant uncertainty, which they referred to as **guaranteed-cost** control. Basically, the idea was to minimize an upper bound on the performance measure and hence guarantee a certain level of performance for all allowable plant perturbations. Unfortunately, the theory required the solution of a complicated matrix-differential equation that was not accessible to simple analysis. In addition, considerable overbounding was necessary to obtain the necessary guarantees on robust stability and performance. Robust performance for the linear-quadratic case still remains an important, open research problem.

In the early eighties, H^∞ space concepts were introduced for the design of robust control systems. Via **Q-parameter** [31] and **U-parameter** [55] methods, this theory was applied to the design of robust LQR and LQG systems.

Finally, it should be noted that by the mid eighties, software for the solution of linear-quadratic problems became commercially available, and numerous applications of the theory, especially to multivariable aerospace problems, had been reported.

1.2 Prerequisites

In the development of the linear-quadratic theory in this text, certain prerequisite material is assumed. In particular, it is assumed that graduate-level courses on state-space and probabilistic methods precede the course in linear-quadratic control. No attempt is made to develop any of this material here; rather, the reader is referred to appropriate texts on the subject. We summarize next the required prerequisites and cite some related texts.

- **Basic feedback concepts**. Closed-loop systems, transfer function and block-diagram analysis, sensitivity function, root-locus analysis, Nyquist stability analysis, basic PID compensation, and so on (see, e.g., Franklin, Powell, and Emami-Naeni [69]).

- **Matrix theory**. Basic matrix algebra, rank and singular matrices, eigenvalues and eigenvectors, similarity transformations, singular values and singular-value decomposition, positive-definite and positive-semidefinite quadratic forms, functions of a matrix, and so on (see, e.g. Gantmacher [74], or Barnett [16]).

- **State-space methods**. State-space modeling, controllability, observability, stability, state-transition matrix, pole placement, and observer design, continuous- and discrete-time systems. (see, e.g., Kailath [99] or Chen [45]).

- **Multivariable calculus**. Gradients, multivariable Taylor expansion, and multivariable minimization (Lagrange multipliers). (see, e.g., Luenberger [124]).

- **Probability theory**. Expected values, variance, conditional probabilities, correlation function, white noise, Markov process,

Wiener process, and covariance matrices. (see, e.g., Maybeck [134]).

1.3 Organization and Scope of Text

As noted in the Preface, this text is meant to fill the gap between chapters devoted to the linear-quadratic problem in various books on optimal and multivariable control (e.g., in Athans and Falb [12], Bryson and Ho [35], Friedland [71], Grimble and Johnson [82], Lewis [122], Maciejowski [126], Patel and Munro [158], etc.,) and specialized, advanced books that deal almost exclusively with the subject (such as Anderson and Moore [4], Kwakernaak and Sivan [113], and Mehrmann [139]).

Although not formally partitioned, the text divides naturally into four parts:

- **Optimal state feedback**. Optimal regulator (LQR) problem (Chapter 2), optimal tracking and disturbance-rejection problem (Chapter 3), and robustness properties (Chapter 4).

- **Stochastic control**. Optimal state-feedback control of stochastic systems (Chapter 5).

- **Optimal output feedback**. Optimal output-feedback control with Gaussian sensor and system disturbance signals, i.e., the LQG problem (Chapter 6).

- **Robust design**. LQG/LTR and other robust design approaches (Chapter 7).

The theory is developed entirely in the continuous-time domain. Given the extent to which digital control is currently used in practice, one could argue that a discrete-time approach would be more appropriate. We have elected to use a continuous-time approach for three main reasons:

1. Control engineers are more familiar with continuous-time-system theory than discrete-time-system theory. Of course, this may change in the future.

2. Most plants are analog, even if the controllers are digital.

3. Digital-signal-processing chips that can emulate analog controllers quite well for most applications, are widely available.

Nevertheless we have included a chapter (Chapter 8) on digital controllers and on the direct design of discrete-time LQR controllers.

We have also included simple scalar examples in various sections of the text. This may seem inconsistent with the main multivariable thrust of the text; however, the purpose of the examples is to illustrate certain aspects of the theory that can be extended to multivariable systems, with problems that can easily be computed by hand. More practical multivariable problems, which almost always require computer solutions, are included in the software sections near the end of various chapters. Also included at the end of most chapters are " homework" problems, and notes and references with comments on the material presented and related references. We have elected to use MATLAB® as the software for linear-quadratic design. It should be noted, however, that many other software packages are also commercially available.

Finally, we wish to emphasize again that this is meant to be an introduction to the subject and not an advanced treatise. Thus, we have often chosen the least complicated cases to discuss, such as square transfer function matrices, and have omitted detailed discussions on a number of worthy topics. Also, to keep the material as friendly as possible to engineers, we have avoided long lists of formal definitions and theorems. Thus, while we have attempted to clearly identify and prove most of the major results, nothing is labeled a theorem.

Chapter 2

The LQR Problem

In this chapter the optimal state-feedback, linear-quadratic-regulator (LQR) problem is formulated and solved. Dynamic programming is used to obtain the necessary optimization equation, and a state-feedback solution is obtained in terms of the solution of a differential, matrix Riccati equation. The steady-state regulator problem is then solved, followed by a discussion of the design issues involved in the choice of the weighting matrices. The chapter concludes with a numerical example to illustrate the LQR solution using MATLAB.

2.1 LQR Problem Statement

The basic optimization problem considered in this chapter is that of finding a state-feedback control law of the form $u = -Kx$ that minimizes a performance measure of the form

$$V = \int_0^T (x'Qx + u'Ru)dt + x'(T)Mx(T) \tag{2.1}$$

where the system dynamics are given by

$$\dot{x} = Ax + Bu \tag{2.2}$$

and Q and M are typically positive-semidefinite matrices, R is a positive-definite matrix, x is an n-dimensional state vector, and u is an m-dimensional input vector. In this problem we consider the final time T fixed, but the final state $x(T)$ free.

It is assumed that the state equations (2.2) have been written so that the state x represents an incremental state. Thus, the control goal is to keep the state x as close to the state $x_r = 0$ as possible. The term $x'Qx$ in (2.1) is then a measure of control accuracy, the term $u'Ru$ is a measure of control effort, and the term $x'(T)Mx(T)$ is a measure of terminal control accuracy. A control problem where the object is to maintain the state close to the zero state is referred to as a **regulator problem;** hence, we refer to the above problem as the optimal **linear-quadratic-regulator (LQR) problem**. Note that problems that require a minimization of control accuracy, as measured for example by

$$V = \int_0^T x'Qx \, dt + x'(T)Mx(T) \tag{2.3}$$

subject to an explicit control-effort constraint of the form

$$\int_0^T u'Ru \, dt = 1 \tag{2.4}$$

can be converted to the form given in (2.1) by multiplying (2.4) by a Lagrangian multiplier λ and adding the result to (2.3).

2.2 Hamilton–Jacobi Equation

In this section we derive an optimization equation for the optimal control problem defined above. Since generalizing the problem to a nonlinear time-varying problem does not complicate the derivation of the optimization equation, we in fact consider the more general problem of finding a state-feedback control law $u = \phi(x, t)$, such that the general performance measure

$$V = \int_t^T l(x, u, \tau) d\tau + m[x(T)] \tag{2.5}$$

is minimized, given the nonlinear system dynamics

$$\dot{x} = f(x, u, t) \tag{2.6}$$

The "loss" function $l(x, u, \tau)$ and terminal cost $m(x)$ are generally non-negative functions of x and u, with $l(0, 0, \tau) = 0$ and $m(0) = 0$. While the generalization to nonlinear time-varying systems causes no complications in deriving the optimization equation, it does cause severe problems in obtaining solutions to the optimization equation.

We next use the **optimality principle** to derive an optimization equation for the above optimal control problem. The basic assumption underlying the optimality principle is that the system can be characterized by its state $x(t)$ at time t, which completely summarizes the effect of all inputs $u(\tau)$ prior to time t. The use of the optimality principle to derive optimization equations for optimal control was first proposed by Bellman [18] and is commonly referred to as the **dynamic-programming approach.**

Optimality Principle. *If $u^*(\tau)$ is optimal over the interval $[t, T]$, starting at state $x(t)$, then $u^*(\tau)$ is necessarily optimal over the subinterval $[t + \Delta t, T]$ for any Δt such that $T - t \geq \Delta t > 0$.*

Proof: (By contradiction.) Assume there exists a u^{**} that yields a smaller value of

$$\int_{t+\Delta t}^{T} l(x, u, \tau) d\tau + m[x(T)] \tag{2.7}$$

than u^* over the subinterval $[t + \Delta t, T]$. Consider now a new control input, $u(\tau)$, given by

$$u(\tau) = \begin{cases} u^*(\tau), & \text{for } t \leq \tau \leq t + \Delta t \\ u^{**}(\tau), & \text{for } t + \Delta t \leq \tau \leq T \end{cases} \tag{2.8}$$

Then over the interval $[t, T]$, we have

$$\begin{aligned} &\int_{t}^{t+\Delta t} l(x^*, u^*, \tau) d\tau + \int_{t+\Delta t}^{T} l(x^{**}, u^{**}, \tau) d\tau + m[x^{**}(T)] \\ &< \int_{t}^{t+\Delta t} l(x^*, u^*, \tau) d\tau + \int_{t+\Delta t}^{T} l(x^*, u^*, \tau) d\tau + m[x^*(T)] \end{aligned} \tag{2.9}$$

But u^* is optimal, by assumption, over the interval $[t, T]$, and (2.9) implies that u given by (2.8) yields a smaller value of performance than the optimal. This is obviously a contradiction. ■

Note that in the above, x^* denotes the state trajectory corresponding to u^*, x^{**} denotes the state trajectory corresponding to u^{**}, and finally $x^* = x^{**}$ at $\tau = t + \Delta t$ since u and u^* are the same over the first interval (t, T).

We next derive an equation, referred to as the **Hamilton–Jacobi optimization equation**, for the above general, optimal control problem. Denote by $V^*(x,t)$ the minimal value of the performance measure V when the initial time is t and the initial state is $x(t)=x$ and denote by $u[t,T]$ the control signal defined over the interval $[t,T]$. Then

$$V^*(x,t) = \min_{u[t,T]} \left\{ \int_t^T l(x,u,\tau)d\tau + m[x(T)] \right\} \tag{2.10}$$

From the additive properties of integrals and the optimality principle we then have

$$V^*(x,t) = \min_{u[t,t+\Delta t]} \left\{ \int_t^{t+\Delta t} l(x,u,\tau)d\tau + V^*[x(t+\Delta t), t+\Delta t] \right\} \tag{2.11}$$

Note that we have used the fact in (2.11) that the minimal value of V starting in state $x(t+\Delta t)$ at time $t+\Delta t$ is given by $V^*[x(t+\Delta t), t+\Delta t]$. *Note also that by using the optimality principle, the problem of finding an optimal control over the interval $[t,T]$ has been reduced to finding an optimal control over the reduced interval $[t, t+\Delta t]$.*

If we now approximate the integral in (2.11) by $l(x,u,t)\Delta t$ and do a multivariable Taylor-series expansion of $V^*[x(t+\Delta t), t+\Delta t]$ about the point $(x(t),t)$, with $x(t+\Delta t)-x(t)$ approximated by $f(x,u,t)\Delta t$, we obtain

$$\begin{aligned} V^*(x,t) &= \min_{u(t)} \{ l(x,u,t)\Delta t + V^*(x,t) + \frac{\partial V^*}{\partial t}\Delta t \\ &\quad + \left[\frac{\partial V^*}{\partial x}\right]' f(x,u,t)\Delta t + o(\Delta t)\} \end{aligned} \tag{2.12}$$

where $\partial V^*/\partial x$ denotes the gradient (a column vector) of V^* with respect to the vector x, and $o(\Delta t)$ denotes higher-order terms in Δt. If we now take the limit as Δt goes to zero and note that $V^*(x,T) = m(x)$ for all x, we obtain the Hamilton-Jacobi optimization equation:

$$\begin{aligned} -\partial V^*/\partial t &= \min_{u(t)} \left\{ l(x,u,t) + \left[\frac{\partial V^*}{\partial x}\right]' f(x,u,t) \right\}; \\ V^*(x,T) &= m(x), \text{ for all } x \end{aligned} \tag{2.13}$$

To obtain a solution to the optimization equation, (2.13), we proceed in two steps. The first is to perform the indicated minimization. This leads to a control law of the form

$$u^* = \psi(\frac{\partial V^*}{\partial x}, x, t). \tag{2.14}$$

The second is to substitute (2.14) back into (2.13) and solve the nonlinear, partial-differential equation

$$-\frac{\partial V^*}{\partial t} = l(x, \psi, t) + \left[\frac{\partial V^*}{\partial x}\right]' f(x, \psi, t) \tag{2.15}$$

for $V^*(x,t)$, subject to the boundary condition

$$V^*(x, T) = m(x) \tag{2.16}$$

We can then compute the gradient of $V^*(x,t)$ with respect to x and hence obtain the optimal state-feedback control law

$$u^* = \psi(\frac{\partial V^*}{\partial x}, x, t) = \phi(x, t) \tag{2.17}$$

In general it is not possible to solve (2.15) analytically. However, as will be shown in the next section, an analytic solution is possible for the LQR problem. Finally, it should be noted that solving the Hamilton-Jacobi optimization equation is only a necessary condition for optimality. The sufficiency of any solution so obtained must be checked separately as discussed in Section 2.6.

2.3 LQR Solution: Matrix Riccati Equation

For the LQR problem we have $l(x, u, t) = x'Qx + u'Ru$, $f(x, u, t) = Ax + Bu$, and $m(x) = x'Mx$. The matrices A, B, Q, R, and M constitute the input data to the LQR problem and in general may all be time-varying, i.e., depend explicitly on t. For convenience, this dependency is suppressed in the sequel. Most of the results in this section are valid for the general time-varying case. Indeed, this is an important advantage of the dynamic-programming solution to linear-quadratic problems, as opposed to frequency-domain approaches. Results that are valid only in the time-invariant case will be explicitly noted.

The first step in obtaining a solution to the LQR problem is the minimization of

$$x'Qx + u'Ru + \left[\frac{\partial V^*}{\partial x}\right]' (Ax + Bu) \tag{2.18}$$

with respect to u. This minimization may be done by setting the gradient of (2.18) to the zero vector. When R is positive definite, this necessary condition for a minimal point is also a sufficient condition. Setting the gradient equal to zero and solving for u yields

$$u^* = -\frac{1}{2}R^{-1}B'\frac{\partial V^*}{\partial x} \tag{2.19}$$

Recall that the gradient of a quadratic formula $u'Wu + b'u$ with respect to u is equal to $2Wu + b$ (see, e.g., [124]). This was used in deriving (2.19) and will be used again shortly in computing the gradient of V^* with respect to x. At this point it would be useful to know the form of V^*. It is known that an integral-quadratic form evaluated for a linear system is a quadratic form in the initial state of the system; hence, it is reasonable to assume that

$$V^*(x,t) = x'P(t)x, \tag{2.20}$$

where $P(t)$ is symmetric. The gradient of V^* is then $2P(t)x$. If V^* given in (2.20) and its gradient are substituted back into (2.15), we obtain after some matrix manipulations

$$-x'\dot{P}x = x'[A'P + PA + Q - PBR^{-1}B'P]x \tag{2.21}$$

with the boundary condition

$$V^*(x,T) = x'P(T)x = x'Mx \tag{2.22}$$

Note that we have used the matrix identity $2x'PAx = x'(A'P+PA)x$ in deriving (2.21). Since (2.21) and (2.22) must be true for all x, we obtain the following **matrix Riccati equation** and final boundary value for $P(t)$

$$\begin{aligned} -\dot{P} &= A'P + PA + Q - PBR^{-1}B'P; \\ P(T) &= M \end{aligned} \tag{2.23}$$

Finally, the optimal state-feedback control law is given by

$$\begin{aligned} u^* &= -K(t)x; \\ K(t) &= R^{-1}B'P(t) \end{aligned} \tag{2.24}$$

Figure 2.1 shows the LQR state-feedback configuration corresponding to the control law $u = -K(t)x$. Note from (2.20) that $P(t)$ also

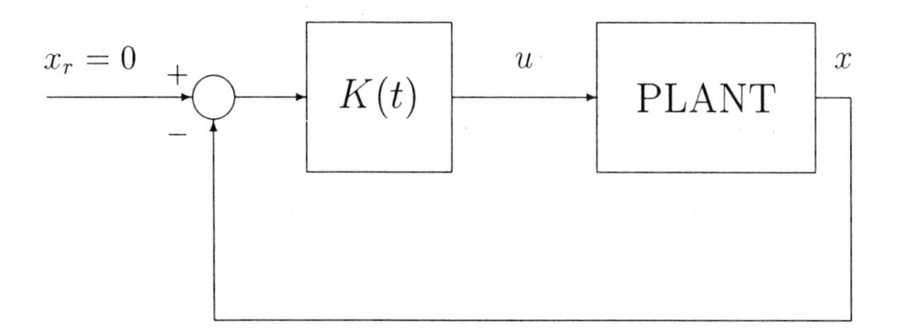

Figure 2.1: Optimal LQR State-Feedback Configuration

provides an evaluation of the optimal performance measure. The problem of solving the matrix Riccati equation (2.23) is nontrivial and will be discussed in the next section. To illustrate some of the properties of the solution, however, we solve next a simple scalar problem.

Example 2.1 Consider the problem of minimizing

$$V = \int_0^T (x^2 + u^2)dt$$

subject to

$$\dot{x} = u$$

Solution: The data in this case are: $A = M = [0], B = Q = R = [1]$. The Riccati equation to be solved then becomes

$$-\dot{P} = 1 - P^2, \quad P(T) = 0$$

This scalar Riccati differential equation may be solved by separation of variables. The solution is

$$P(t) = \frac{1 - e^{-2(T-t)}}{1 + e^{-2(T-t)}}$$

with optimal control

$$u^*(t) = -P(t)x(t)$$

Note that in the above example the control law is time-varying even though all the original data were time-invariant. However, if the optimization interval becomes infinite, i.e., the "time-to-go" $T-t$ approaches infinity, the control law becomes time-invariant, i.e., $K(t)$ approaches the constant one. Finally, note that in the steady state (infinite time-to-go), the optimal closed-loop system is asymptotically stable, while the open-loop system ($u = 0$) was not asymptotically stable. With appropriate assumptions, these properties of the optimal LQR solution hold even in the vector case.

2.4 Riccati Equation Solution

We show next that the (nonlinear) Riccati equation can be solved by solving the following system of **linear** *equations*

$$\begin{bmatrix} \dot{X} \\ \dot{Y} \end{bmatrix} = \begin{bmatrix} A & -BR^{-1}B' \\ -Q & -A' \end{bmatrix} \begin{bmatrix} X \\ Y \end{bmatrix} \tag{2.25}$$

where X and Y are $n \times n$ matrices that satisfy the following boundary conditions

$$\begin{bmatrix} X(T) \\ Y(T) \end{bmatrix} = \begin{bmatrix} I \\ M \end{bmatrix} \tag{2.26}$$

The solution of the Riccati equation (2.23), $P(t)$, is then given by

$$P(t) = Y(t)X^{-1}(t) \tag{2.27}$$

The matrix

$$H = \begin{bmatrix} A & -BR^{-1}B' \\ -Q & -A' \end{bmatrix} \tag{2.28}$$

which appears in (2.25), plays an important role in linear-quadratic problems and is called the **Hamiltonian matrix**.

Proof: Note that for any invertible matrix X

$$\frac{dYX^{-1}}{dt} = Y\frac{dX^{-1}}{dt} + \frac{dY}{dt}X^{-1} \tag{2.29}$$

Now from

$$X(t)X^{-1}(t) = I$$

we obtain, by differentiating both sides with respect to t,

$$\frac{dX^{-1}}{dt} = -X^{-1}\frac{dX}{dt}X^{-1} \tag{2.30}$$

If we substitute (2.30) and the derivatives of X and Y given in (2.25) into (2.29) we obtain

$$\frac{d(YX^{-1})}{dt} = -A'YX^{-1} - YX^{-1}A - Q + YX^{-1}BR^{-1}B'YX^{-1} \tag{2.31}$$

With the substitution $P = YX^{-1}$ into (2.31) it is clear that YX^{-1} satisfies the Riccati equation (2.23). ■

When the matrices A, B, Q, and R are time-varying, the linear equation (2.25) is also time-varying and analytical solutions are not generally available. When the matrices are time-invariant (constant), however, an explicit solution is available in terms of matrix exponential functions. We outline next the procedure for obtaining the solution in this case.

We first note the fact that if λ is an eigenvalue of the Hamiltonian matrix H, then so is $-\lambda$. This may be easily shown by noting that the matrix H' is mathematically similar to the matrix $-H$ and thus has the same eigenvalues, that there is a nonsingular transformation J such that $-H = JH'J^{-1}$. The required J is just

$$J = \begin{bmatrix} 0 & I \\ -I & 0 \end{bmatrix}$$

Let U be a nonsingular transformation such that

$$U^{-1}HU = \begin{bmatrix} \Lambda_s & 0 \\ 0 & \Lambda_u \end{bmatrix}$$

where Λ_s is an $n \times n$ Jordan block matrix with all eigenvalues having negative real parts (stable matrix), and Λ_u is an $n \times n$ Jordan block matrix with all eigenvalues having positive real parts (completely unstable matrix). If H has no purely imaginary eigenvalues such a transformation always exists. Conditions for the exclusion of such eigenvalues will be discussed subsequently. Furthermore, let U be partitioned into four $n \times n$ blocks

$$U = \begin{bmatrix} U_{11} & U_{12} \\ U_{21} & U_{22} \end{bmatrix}$$

where

$$\begin{bmatrix} U_{11} \\ U_{21} \end{bmatrix} \tag{2.32}$$

has columns made up of generalized eigenvectors of H corresponding to eigenvalues with negative real parts, and where

$$\begin{bmatrix} U_{12} \\ U_{22} \end{bmatrix} \tag{2.33}$$

has columns made up of generalized eigenvectors of H corresponding to eigenvalues with positive real parts. Now with the transformation

$$\begin{bmatrix} X \\ Y \end{bmatrix} = U \begin{bmatrix} \hat{X} \\ \hat{Y} \end{bmatrix} \tag{2.34}$$

the differential equation (2.25) becomes

$$\frac{d}{dt}\begin{bmatrix} \hat{X} \\ \hat{Y} \end{bmatrix} = \begin{bmatrix} \Lambda_s & 0 \\ 0 & \Lambda_u \end{bmatrix}\begin{bmatrix} \hat{X} \\ \hat{Y} \end{bmatrix} \tag{2.35}$$

From the decoupled equations (2.35) one may compute $(\hat{X}, \hat{Y})$ at time T in terms of time t as follows

$$\begin{aligned} \hat{X}(T) &= e^{\Lambda_s(T-t)}\hat{X}(t) \\ \hat{Y}(T) &= e^{\Lambda_u(T-t)}\hat{Y}(t) \end{aligned} \tag{2.36}$$

Evaluating (2.34) at the boundary, $t = T$, one obtains

$$\begin{aligned} I &= U_{11}\hat{X}(T) + U_{12}\hat{Y}(T) \\ M &= U_{21}\hat{X}(T) + U_{22}\hat{Y}(T) \end{aligned} \tag{2.37}$$

Using (2.37) we can solve for $\hat{Y}(T)$ in terms of $\hat{X}(T)$, i.e., $\hat{Y}(T) = G\hat{X}(T)$ where

$$G = -[U_{22} - MU_{12}]^{-1}[U_{21} - MU_{11}] \tag{2.38}$$

Then from (2.34) evaluated at time t we obtain, after some factoring

$$\begin{aligned} X(t) = {} & [U_{11} + U_{12}e^{-\Lambda_u(T-t)}Ge^{\Lambda_s(T-t)}]e^{-\Lambda_s(T-t)}\hat{X}(T) \\ Y(t) = {} & [U_{21} + U_{22}e^{-\Lambda_u(T-t)}Ge^{\Lambda_s(T-t)}]e^{-\Lambda_s(T-t)}\hat{X}(T) \end{aligned} \tag{2.39}$$

If we now compute $P(t) = Y(t)X^{-1}(t)$, canceling the common term

$$e^{-\Lambda_s(T-t)}\hat{X}(T)$$

we finally have

$$\begin{aligned} P(t) &= P_1(t)P_2^{-1}(t) \\ P_1(t) &= [U_{21} + U_{22}e^{-\Lambda_u(T-t)}Ge^{\Lambda_s(T-t)}] \\ P_2(t) &= [U_{11} + U_{12}e^{-\Lambda_u(T-t)}Ge^{\Lambda_s(T-t)}] \end{aligned} \tag{2.40}$$

If the linear equation (2.25) is time-varying, it is possible to express the solutions for $X(t)$ and $Y(t)$ in terms of its transition matrix, $\Phi(t,T)$. Unfortunately, as previously noted, the transition matrix in the time-varying case cannot, in general, be computed analytically. Nevertheless, if the $2n \times 2n$ transition matrix is partitioned into four equal-sized blocks, i.e.,

$$\Phi(t,T) = \begin{bmatrix} \Phi_{11}(t,T) & \Phi_{12}(t,T) \\ \Phi_{21}(t,T) & \Phi_{22}(t,T) \end{bmatrix} \tag{2.41}$$

and boundary condition (2.26) is used, we can compute $X(t)$ and $Y(t)$, i.e.,

$$\begin{aligned} X(t) &= \Phi_{11}(t,T)X(T) + \Phi_{12}(t,T)Y(T) \\ &= \Phi_{11}(t,T) + \Phi_{12}(t,T)M \\ Y(t) &= \Phi_{21}(t,T)X(T) + \Phi_{22}(t,T)Y(T) \\ &= \Phi_{21}(t,T) + \Phi_{22}(t,T)M \end{aligned} \tag{2.42}$$

Then the solution for $P(t)$ becomes

$$P(t) = [\Phi_{21}(t,T) + \Phi_{22}(t,T)M]\,[\Phi_{11}(t,T) + \Phi_{12}(t,T)M]^{-1} \tag{2.43}$$

Example 2.2 Consider again the problem of minimizing

$$V = \int_0^T (x^2 + u^2)dt$$

subject to

$$\dot{x} = u$$

but now using the solution given by (2.40).

Solution: The Hamiltonian matrix for this problem is

$$H = \begin{bmatrix} 0 & -1 \\ -1 & 0 \end{bmatrix}$$

The U matrix is

$$U = \begin{bmatrix} 1 & 1 \\ 1 & -1 \end{bmatrix}$$

where the first column in U is the eigenvector corresponding to the stable eigenvalue, $\Lambda_s = -1$, and the second column is the eigenvector corresponding to the unstable eigenvalue, $\Lambda_u = 1$. If these values are substituted back into (2.40) we obtain, as before,

$$P(t) = \frac{1 - e^{-2(T-t)}}{1 + e^{-2(T-t)}}$$

2.5 Steady-State Regulator Problem

In this section we explore the LQR problem when the data is time-invariant and the optimization interval is infinite, i.e., T approaches infinity. We refer to this special problem as the **steady-state LQR problem**.

One may note immediately from the Riccati solution given by (2.40) that when a steady-state solution exists, the optimal control is time-invariant and given by $u^*(t) = -Kx(t)$, where

$$K = R^{-1}B'U_{21}U_{11}^{-1} \tag{2.44}$$

where U_{21} and U_{11} are defined in (2.32). Denote by $\bar{P}$ the steady-state solution for $P(t)$, then $\bar{P} = U_{21}U_{11}^{-1}$ must satisfy the **algebraic Riccati equation (ARE)**

$$0 = A'\bar{P} + \bar{P}A + Q - \bar{P}BR^{-1}B'\bar{P} \tag{2.45}$$

and the optimal control is given by

$$u^*(t) = -R^{-1}B'\bar{P}x(t) \tag{2.46}$$

The above results follow directly from the fact Λ_s and $-\Lambda_u$ both have eigenvalues with negative real parts, so that the matrix-exponential terms that appear in (2.40) go to the zero matrices as T, or equivalently the time-to-go, $T - t$, goes to infinity. The control law given by (2.46) constitutes the solution to the steady-state LQR problem. Note that the steady-state solution is independent of the matrix M. We discuss next some of the conditions required for the existence of the steady-state solution, and for the stability properties of the resulting closed-loop system. We need to recall first some basic results from state-space theory. In particular:

- The linear system $\dot{x} = Ax + Bu$, $y = Cx$ is **controllable** if and only if

$$\text{rank}[B|AB|\ldots|A^{n-1}B] = n$$

and **observable** if and only if

$$\text{rank}\begin{bmatrix} C \\ CA \\ \vdots \\ CA^{n-1} \end{bmatrix} = n$$

Also, the system is said to be **stabilizable** if the uncontrollable modes are stable, and **detectable** if the unstable modes are observable. For convenience we speak of the pair (A, B), rather than the system, being controllable or stabilizable. Similarly, we speak of the pair (A, C) being observable or detectable.

- If a system is controllable, or stabilizable, there always exists a constant control law $u(t) = -Kx(t)$ that makes the closed-loop system asymptotically stable.

- If the system is observable, then $Ce^{At}x = 0$ implies that $x = 0$.

- The equilibrium point $x = 0$ is asymptotically stable for the linear system $\dot{x} = Ax$ if there exists a **Lyapunov function** $V = x'Px$, where $P > 0$ and such that $\dot{V} \leq 0$; and where $\dot{V} \equiv 0$ implies $x(t) \equiv 0$.

The above results may be found in most texts on linear system theory (see, e.g., Kailath [99]). We proceed next to prove the following key result:

Existence and Stability of the Steady-State LQR Solution. *Given the LQR problem with $M = 0$, $R > 0$, $Q = D'D$, where the pair (A, D) is observable and the pair (A, B) is controllable, it follows that a solution to the steady-state LQR problem exists, in particular, that there exists a unique positive-definite solution to the ARE (2.45), $\bar{P}$, and that the optimal closed-loop system, i.e., $\dot{x} = (A - BK)x$, where $K = R^{-1}B'\bar{P}$, is asymptotically stable.*

Proof: A few steps are required to prove the above results. First, from the assumptions that $M = 0, R > 0$, and $Q = D'D$ (the matrix

D is generally such that rank Q = number of rows of D), and the fact that $V^*(x,t) = x'\bar{P}x$, for $T = \infty$, it follows that $\bar{P}$ is bounded below (in the quadratic-form sense) by the zero matrix, i.e., $0 \leq \bar{P}$. *Next we show that* $P(t)$ *is bounded above for all* T, *and that* $P(t)$ *is monotonically increasing with increasing* T. In fact, this is sufficient to prove that a limiting solution to the Riccati equation (2.23) exists, since a monotonically increasing sequence that is bounded above is known to converge. To show that $P(t)$ is monotonically increasing with T, consider the inequality

$$\int_t^{T_1} l(x^*, u^*, \tau)d\tau \leq \int_t^{T_2} l(x^*, u^*, \tau)d\tau = V^*(T_2)$$

where $T_2 \geq T_1$. The inequality follows directly from the additive property of integrals and the non-negativity of $l(x,u,t) = x'Qx + u'Ru$. Here u^* denotes the control input that is optimal over the interval $[t, T_2]$, with corresponding state $x^*(t)$. We also let $V^*(T)$ denote for the moment the optimal value of V over the interval $[t, T]$. Then if u^{**} denotes the control input that is optimal over the interval $[t, T_1]$ we have another inequality

$$V^*(T_1) = \int_t^{T_1} l(x^{**}, u^{**}, \tau)d\tau \leq \int_t^{T_1} l(x^*, u^*, \tau)d\tau$$

since u^{**} always yields a value of V, that is less than that of u^* over the subinterval (t, T_1). It then follows that

$$V^*(T_1) \leq V^*(T_2), \quad \text{for all } T_1 \leq T_2$$

which implies, since V^* is quadratic in P, that $P(t)$ is monotonically increasing with respect to T as required.

We next show that $P(t)$ is bounded above for all T. From the controllability assumption it follows that there exists a constant matrix K, not necessarily optimal, such that the closed-loop system $\dot{x} = (A - BK)x$ is asymptotically stable. This is the standard result from state-variable theory quoted above. The value of V corresponding to the control law $u(t) = -Kx(t)$ is then finite, since $x(t)$ and $u(t)$ are exponentially bounded. Let this value, necessarily quadratic in x, be denoted $x'\hat{P}x$. Then for any $V^*(x,t) = x'P(t)x$ that is optimal over (t, ∞) one must have

$$P(t) \leq \hat{P}$$

This establishes the required upper bound on $P(t)$.

Now we show that $\bar{P}$ must be positive-definite if the pair (A, D) is observable. Assume $\bar{P}$ is only semidefinite. We will show this leads to a contradiction. If $\bar{P}$ is only positive-semidefinite, there exists a nonzero $x(0)$ such that $x'(0)\bar{P}x(0) = 0$, i.e.,

$$x'(0)\bar{P}x(0) = \int_0^\infty [x'(t)D'Dx(t) + u'(t)Ru(t)]dt = 0.$$

Since $R > 0$, this implies that $u(t)$ is identically zero over the interval $[0, \infty]$, and hence that

$$\int_0^\infty [(Dx(t))'Dx(t)]dt = 0$$

and this in turn implies that $De^{At}x(0) = 0$ over the interval $[0, \infty]$. From the observability assumption this last equality then implies that $x(0) = 0$, a contradiction with the hypothesis that $x(0)$ was nonzero.

We next prove that with all the above assumptions, the optimal closed-loop system is asymptotically stable. To this end note that the ARE (2.45) can be rewritten

$$A_c'\bar{P} + \bar{P}A_c = -D'D - \bar{P}BR^{-1}B'\bar{P}$$

where A_c is the closed-loop system matrix given by

$$A_c = A - BR^{-1}B'\bar{P}$$

From the fact that $\bar{P}$ is known to be positive definite and the pair (A, D) is observable by assumption, it follows from the Lyapunov-function stability result quoted above that the matrix A_c must be a stability matrix. To prove this consider the Lyapunov function $V = x'\bar{P}x$, where $\bar{P}$ is the solution of the ARE. Then $\dot{V} = -x'D'Dx - x'(R^{-1}B'\bar{P})'R(R^{-1}B'\bar{P})x$ and $\dot{V} \equiv 0$ implies that $R^{-1}B'\bar{P}x(t) \equiv 0$, since $R > 0$. This in turn implies that $Dx(t) \equiv 0$ for the system $\dot{x} = Ax$. Then from the observability of the pair (A, D) it follows that $x(0) = 0$ and hence $x(t) \equiv 0$. We can now invoke the Lyapunov-function result quoted above.

The last fact is that the positive-definite solution to the ARE is unique. We will not prove this result here. See [4], p. 49, for an outline of the proof. ■

Actually, the existence and stability results proven above are valid under the weaker conditions of stabilizability and detectability. Without going into details, stabilizability is really all that is needed to guarantee an upper bound on $P(t)$, and detectability is all that is needed to guarantee that A_c is stable. Actually stabilizability and detectability are also necessary conditions for both existence and stabilization. With the condition of observability replaced by detectability, however, the matrix $\bar{P}$ may only be positive semidefinite. *Finally, it can be shown that stabilizability and detectability are all that are required from the system dynamics for the Hamiltonian matrix H to have no purely imaginary eigenvalues.* A proof of this is outlined on p. 50 of [4]. The following two simple examples illustrate some of these points.

Example 2.3 Consider the performance measure

$$V = \int_0^\infty u^2 dt$$

with scalar system dynamics

$$\dot{x} = x + u$$

The optimal solution exists and is obviously $u^* = 0$, but the closed-loop system is unstable, i.e., $\dot{x} = x$. This system is controllable but neither observable nor detectable (the unstable mode e^t is not observable). Recall that controllability is a sufficient condition for a solution to exist and that detectability is the weakest condition for stability of the optimal closed-loop system.

△

Example 2.4 Consider the same performance measure as in the above example but with the system dynamics

$$\dot{x} = -x + u$$

This system is controllable (hence stabilizable) and detectable (since there are no unstable modes). Again, the optimal solution is obviously $u^* = 0$, with $\bar{P} = 0$. Note that the closed-loop system $\dot{x} = -x$, is stable, as expected from the stabilizability and detectability of the dynamical system, but that $\bar{P}$ is not positive definite.

One final result that can be useful is that the closed-loop eigenvalues are precisely the stable eigenvalues of the Hamiltonian matrix, given by (2.28). To prove this note that the matrix H, given by (2.28), is similar to the block upper-triangular matrix

$$W = \begin{bmatrix} A_c & X \\ 0 & -A'_c \end{bmatrix}$$

since $W = T^{-1}HT$ for the matrix

$$T = \begin{bmatrix} I & 0 \\ \bar{P} & I \end{bmatrix}$$

where $\bar{P}$ is the solution of the ARE (2.45). Note that T^{-1} is given by

$$T^{-1} = \begin{bmatrix} I & 0 \\ -\bar{P} & I \end{bmatrix}$$

and that the ARE is used in zeroing out the $(2,1)$ entry in W.

2.6 Sufficient Conditions for Optimality

As previously noted, the Hamilton–Jacobi equations were derived as only necessary conditions for optimality. *It can be shown, however, that when the performance measure V can be evaluated for a given feedback-control law, the function V is continuous in its arguments, and all the necessary derivatives exist, then the satisfaction of the inequality*

$$l(x,\phi,t) + \left[\frac{\partial V^{\phi}}{\partial x}\right]' f(x,\phi,t) \leq l(x,u,t) + \left[\frac{\partial V^{\phi}}{\partial x}\right]' f(x,u,t), \quad (2.47)$$

for all x and u, where $V^{\phi}(x,t)$ represents the value of performance corresponding to the state-feedback control law $u^* = \phi(x,t)$, is sufficient to guarantee that the control law u^* is optimal.

Proof: First of all note that

$$V^{\phi}(x,t) = \int_t^T l[x^*(\tau),\phi(\tau),\tau]d\tau + m(x^*(T)) \quad (2.48)$$

where $x^*(\tau)$ is the state corresponding to the control $u^* = \phi$ and x is the initial state $x(t) = x$. Next note that V^ϕ is given by (2.48) if and only if it satisfies the equation

$$\begin{aligned} -\frac{\partial V^\phi}{\partial t} &= l(x, \phi, t) + \left[\frac{\partial V^\phi}{\partial x}\right]' f(x, \phi, t), \\ V^\phi(x, T) &= m(x), \text{ for all } x \end{aligned} \tag{2.49}$$

The fact that (2.48) implies (2.49) follows directly from the same kind of argument used to derive the Hamilton–Jacobi optimization equation. The converse can be shown by integrating both sides of (2.49) over the interval $[t, T]$, noting that

$$\begin{aligned} \int_t^T \frac{dV^\phi}{d\tau} d\tau &= \int_t^T [\frac{\partial V^\phi}{\partial \tau} + \left[\frac{\partial V^\phi}{\partial x}\right]' f(x, \phi, \tau)] d\tau \\ &= V^\phi(x(T), T) - V^\phi(x(t), t) \end{aligned} \tag{2.50}$$

and noting that $V^\phi(x, T) = m(x)$ for all x.

To prove that (2.47) is a sufficient condition for optimality, add the term $\partial V^\phi / \partial t$ to both sides of this equation and integrate from t to T. Then use the result in (2.50) to obtain

$$0 \leq V^\phi(x(T), T) - V^\phi(x(t), t) + \int_t^T l[x(\tau), u(\tau), \tau] d\tau \tag{2.51}$$

where $x(\tau)$ denotes the trajectory corresponding to the control input $u(\tau)$. Using the fact that $V^\phi(x, T) = m(x)$, one obtains the inequality

$$V^\phi(x, t) \leq \int_t^T l[x(\tau), u(\tau), \tau] d\tau + m(x(T)) \tag{2.52}$$

Inequality (2.52) states that the performance measure corresponding to the control law $u^* = \phi(x, t)$ yields a value that is less than or equal to that obtainable with any other control law u, thus establishing the optimality of u^*. ■

We can now show, for example, that the steady-state solution obtained in Section 2.5 is indeed optimal. Since $V^* = x'\bar{P}x$ and u^* given by $u^* = -R^{-1}B'\bar{P}x$ minimizes

$$x'Qx + u'Ru + 2x'\bar{P}(Ax + Bu) \tag{2.53}$$

it follows that the inequality (2.47) is satisfied for all x and u.

We conclude this section with a method of obtaining a sequence of "improved" control laws, referred to as the method of **approximation-in-policy-space** which is based on the equalities and inequalities developed above. In particular we have the following result:

Approximation-in-Policy-Space Algorithm. *If $V^i(x,t)$ is the performance value computed for the control law u^i and the following inequality is true for all $x \neq 0$*

$$l(x, u^{i+1}, t) + \left[\frac{\partial V^i}{\partial x}\right]' f(x, u^{i+1}, t) < l(x, u^i, t) + \left[\frac{\partial V^i}{\partial x}\right]' f(x, u^i, t) \tag{2.54}$$

then the control law u^{i+1} yields an improved value of performance, i.e., $V^{i+1}(x,t) < V^i(x,t)$ for all nontrivial initial states $x(t) = x$.

Proof: The proof of this result follows the same arguments as above, i.e., add $\partial V^i/\partial t$ to both sides of (2.54), integrate from t to T, etc.

■

When the approximation-in-policy-space algorithm is applied to the steady-state LQR problem one obtains the following algorithm for computing a sequence of monotonically improved control laws:

LQR Approximation-in-Policy-Space. Given any feedback matrix K_i that makes $A - BK_i$ stable, solve the Lyapunov equation

$$(A - BK_i)'P_i + P_i(A - BK_i) = -(Q + K_i'RK_i) \tag{2.55}$$

for P_i, then select u_{i+1} to minimize

$$u'Ru + \left[\frac{\partial V^i}{\partial x}\right]' (Ax + Bu) \tag{2.56}$$

where $V^i = x'P_ix$. This last step yields the improved control law

$$u_{i+1} = -K_{i+1}x = -R^{-1}B'P_ix \tag{2.57}$$

If $(A - BK_{i+1})$ is stable, increase i by one and repeat the above procedure. This is essentially the iterative procedure reported in [108], where it is shown that with the usual assumptions of controllability and observability the procedure does lead to a sequence that converges to the optimal solution.

2.7 Selection of M, Q, and R Matrices

We have assumed so far that the LQR performance-measure matrices are given. It may be that in some problems the choice of these matrices is transparent. Indeed, we will discuss in this section three such special problems, the **cheap-control problem**, the **terminal-control problem**, and the **degree-of-stability problem**. In general, however, it is difficult to translate time- and frequency-domain specifications into specific values of LQR performance-measure matrices. Some rules have been proposed (see e.g., [4]), but most results are only approximate at best. A simple guideline is to select these matrices to be diagonal and to make the diagonal entry "positive large" for any variable we would like to be "small" in the time domain. Of course, this is a rather crude rule, and some further tuning of matrix values will no doubt be required. However, an important advantage of the optimal LQR solution is that any tuning of the performance-measure matrices, within the constraint of positive-semidefiniteness/positive-definiteness, will preserve **asymptotic stability** and **robustness** (see Chapter 4) even if LQ optimality is sacrificed. In any case, we discuss next the three above-mentioned problems.

Cheap Control. By "cheap control" we mean a problem where the performance measure is given by

$$V = \int_t^T (x'Qx + \rho u'Ru)d\tau + x'(T)Mx(T) \tag{2.58}$$

where ρ is a positive scalar that approaches zero.

The term "cheap" refers to the fact that the control effort is viewed as being inexpensive and discounted by the small values of ρ. A major question that arises in this case is, Is it possible, given that the control effort is arbitrarily "cheap," to make the optimal performance value approach zero? The answer is yes, at least in the case where the matrix $D(sI - A)^{-1}B$ has all its finite zeros inside the left-half s-plane (where $Q = D'D$ and D has the same number of rows as the number of columns of B). The zeros of the square matrix $D(sI - A)^{-1}B$ are defined simply as the poles of its inverse (see sect. 3.8.3 of [113] for more details on cheap control). A proof of this result is deferred until Chapter 4, where the **Kalman identity** is derived and applied to this problem. It should be noted that the problem of cheap control can be related to the problem of "expensive-control" accuracy, where the weight of Q is arbitrarily increased by simply dividing the V given in (2.58) by ρ.

Terminal Control. By a "terminal-control" problem we mean one where the terminal matrix is given by ρM, with M positive definite, $Q = 0$, and ρ is allowed to approach ∞. Thus, in the terminal-control problem the control accuracy is penalized only at the terminal time ($Q = 0$) and the terminal state is forced in the limit ($\rho = \infty$) to the zero state. It turns out that in this case the Riccati equation, (2.23), has a simple solution, which we demonstrate next. First note that with $Q = 0$, the Riccati equation becomes

$$\begin{aligned} -\dot{P} &= A'P + PA - PBR^{-1}B'P, \\ P(T) &= \rho M \end{aligned} \tag{2.59}$$

If the differential equation in (2.59) is multiplied on the right and left by P^{-1} and use is made of the identity

$$\frac{dP^{-1}}{dt} = -P^{-1}\dot{P}P^{-1}$$

the nonlinear Riccati equation reduces to the linear Lyapunov equation

$$\dot{S} = SA' + AS - BR^{-1}B' \tag{2.60}$$

where $S = P^{-1}$ and boundary condition $S(T) = 0$ as ρ approaches ∞.

Degree-of-Stability Design. It is sometimes desired to have all the eigenvalues of the closed-loop system with real parts less than some negative number $-\alpha$, $\alpha > 0$. This is what is commonly referred to as "degree of stability $-\alpha$." It turns out that this is simple to design with LQR theory, with proper time-varying choices of Q and R. In particular, consider the performance measure

$$V = \int_0^\infty e^{2\alpha t}(x'\bar{Q}x + u'\bar{R}u)dt \tag{2.61}$$

where $\bar{Q}$ and $\bar{R}$ are constant matrices. Introduce the variables $\xi(t) = e^{\alpha t}x(t)$ and $v(t) = e^{\alpha t}u(t)$. Then the usual state equation, $\dot{x} = Ax + Bu$, becomes

$$\dot{\xi} = (A + \alpha I)\xi + Bv \tag{2.62}$$

and the performance index (2.61) becomes

$$V = \int_0^\infty (\xi' \bar{Q} \xi + v' \bar{R} v) dt \tag{2.63}$$

With the usual assumptions on the data, the constant LQR problem represented by (2.62) and (2.63) will yield an asymptotically stable $\xi(t)$, and since $x(t) = e^{-\alpha t}\xi(t)$, it follows that $x(t)$ must be asymptotically stable with stability degree equal to $-\alpha$. Thus, to design for degree of stability $-\alpha$, one need only solve the modified ARE

$$0 = (A + \alpha I)' \bar{P} + \bar{P}(A + \alpha I) + \bar{Q} - \bar{P} B \bar{R}^{-1} B' \bar{P} \tag{2.64}$$

and use

$$u^*(t) = -\bar{R}^{-1} \bar{B}' \bar{P} x(t) \tag{2.65}$$

If, however, Q and R are fixed a priori, the problem of minimizing an integral-quadratic performance, subject to constraints on the locations of the closed-loop eigenvalues, is a more difficult optimization problem. We conclude this section with a simple example of terminal control.

Example 2.5 Consider the problem of minimizing

$$V = \int_0^T u^2(t) dt + \rho x^2(T)$$

subject to

$$\dot{x} = u$$

with ρ approaching ∞.

Solution: In this case, the equation for $S = P^{-1}$ becomes

$$\dot{S} = -1, \text{ with } S(T) = 0$$

The solution of this equation can be obtained by simple integration. In particular

$$S(t) = T - t$$

with a corresponding optimal control law given by

$$u^*(t) = -R^{-1} B' P(t) x(t) = -(T - t)^{-1} x(t)$$

Note that in this case $x(T) = 0$, but that the feedback gain approaches ∞ as t approaches T.

2.8 MATLAB Software

The MATLAB software package, **Control System Toolbox**, [80], contains many subroutines that compute the optimal feedback-gain matrix K and the Riccati-equation solution P for the steady-state LQR problem (see also [132] for plotting and other numerical routines). One simply enters the matrices $Q, R, A,$ and B and then calls the function **lqr**. In particular,

$$K = lqr(A, B, Q, R)$$

will return the optimal feedback matrix K for the given data $A, B, Q,$ and R, while

$$[K, P] = lqr(A, B, Q, R)$$

will return both K and P. This software solves the problem using the eigenvalue-eigenvector method discussed in Section 2.4. Actually, a Schur decomposition of the H matrix produces more numerically robust results and is used in another routine **lqr2** (see [115] for more details on this approach). Other related routines, include **reg**, **are**, **ric**, and **lqry**. More information on these routines may be obtained online by typing **help** followed by the routine name.

For the case of a finite optimization interval, the time-varying feedback matrix $K(t)$, given by (2.24), may be computed indirectly from the MATLAB functions **impulse** and **eig**. The function **impulse** may be used to compute the matrix exponential required in (2.40), while **eig** may be used to compute the eigenvalues and eigenvectors. It should be noted, however, that the MATLAB software will not compute generalized eigenvectors. In the following, we present an example to illustrate the use of MATLAB in LQR design.

Example 2.6 Aircraft Model This model appeared in Appendix F of [91] and describes a linearized model of the vertical-plane dynamics of an airplane. The incremental-state equations are given by

$$\begin{aligned} \dot{x} &= Ax + Bu \\ y &= Cx \end{aligned}$$

where

$$A = \begin{bmatrix} 0 & 0 & 1.132 & 0 & -1 \\ 0 & -0.0538 & -0.1712 & 0 & 0.0705 \\ 0 & 0 & 0 & 1 & 0 \\ 0 & 0.0485 & 0 & -0.8556 & -1.013 \\ 0 & -0.2909 & 0 & 1.0532 & -0.6859 \end{bmatrix}$$

$$B = \begin{bmatrix} 0 & 0 & 0 \\ -0.12 & 1 & 0 \\ 0 & 0 & 0 \\ 4.419 & 0 & -1.665 \\ 1.575 & 0 & -0.0732 \end{bmatrix} ; \; C = \begin{bmatrix} 1 & 0 & 0 & 0 & 0 \\ 0 & 1 & 0 & 0 & 0 \\ 0 & 0 & 1 & 0 & 0 \end{bmatrix}$$

The states, all incremental, are

- x_1 is the altitude in meters (m)
- x_2 is the forward speed in meters per second (m/s)
- x_3 is the pitch angle in degrees (deg)
- x_4 is the pitch rate in degrees per second (deg/s)
- x_5 is the vertical speed in meters per second (m/s)

and the inputs are

- u_1 is the spoiler angle in tenths of a degree (deg/10)
- u_2 is the forward acceleration in meters per second squared (m/s/s)
- u_3 is the elevator angle in degrees (deg)

and let $Q = C'C$ and $R = I_{3\times 3}$, and the initial-condition vector be $x(0) = [10 \;\; 100 \;\; -15 \;\; 1 \;\; 25]^T$. The command

$$[K, P] = lqr(A, B, Q, R)$$

gives the following gain matrix

$$K = \begin{bmatrix} -0.2642 & -0.0329 & 0.9009 & 0.4240 & -0.1380 \\ 0.3003 & 1.0474 & 0.7179 & 0.1854 & -0.4613 \\ -0.9165 & -0.2750 & -2.5510 & -0.6664 & 1.4070 \end{bmatrix} \tag{2.66}$$

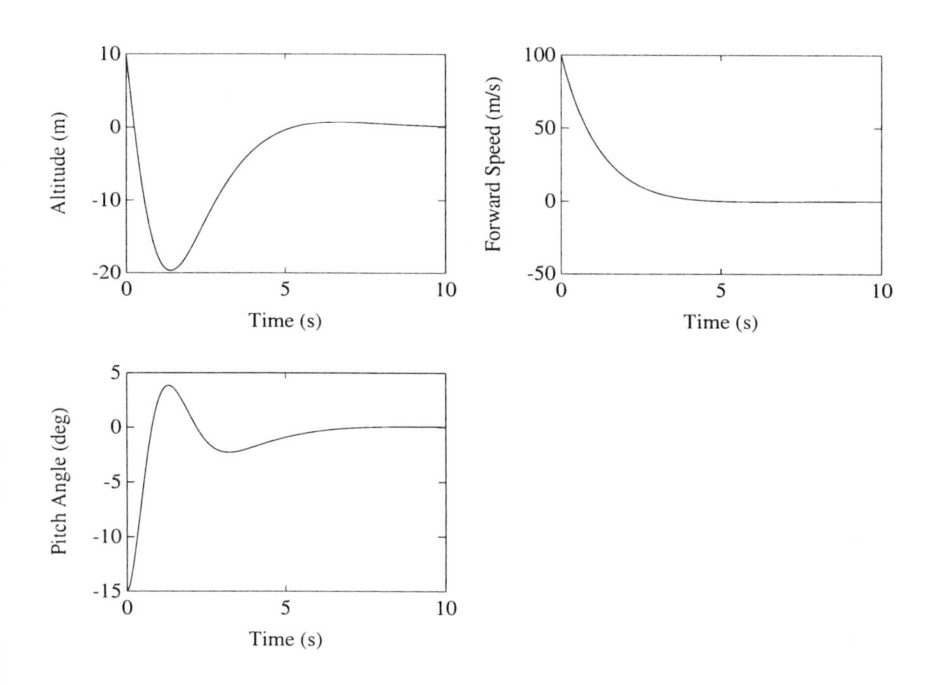

Figure 2.2: Outputs for the Aircraft, $Q = C'C$, $R = I_{3\times3}$

so that $u = -Kx$ will yield the trajectories in Figure 2.2. Examining these figures, we notice that the first output $y_1 = x_1$ is slow to converge to its final value and has a large overshoot. We thus change the matrix Q to have an entry $q_{11} = 100$ while leaving all other entries the same. The results are then illustrated in Figure 2.3. We can see that y_1 converges faster and with less overshoot to its final value since the cost associated with a nonzero x_1 is made larger. Let us next illustrate the effect of increasing the input weighting for the first component of the input vector by letting $r_{11} = 100$ while leaving $q_{11} = 100$. The results are illustrated in Figure 2.4. Notice that the transient behavior of the states is much more pronounced now that we are requiring a smaller input effort.

2.9 Notes and References

We have elected, as in Anderson and Moore [4], to derive the optimization equations for the LQR problem using dynamic programming. However, optimization equations may also be derived from

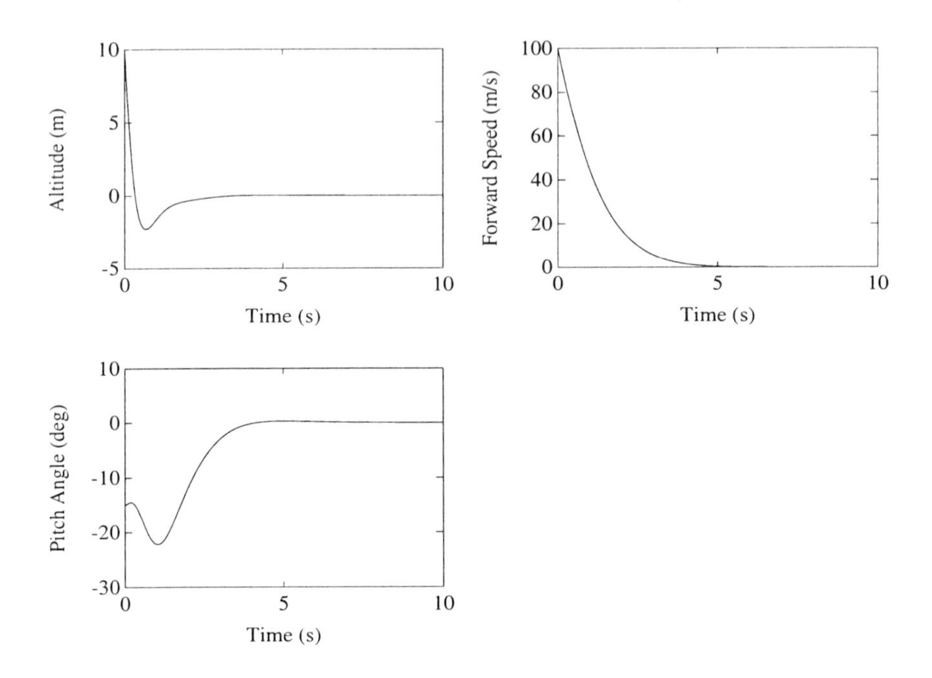

Figure 2.3: Outputs for the Aircraft, $q_{11} = 100$, and $R = I_{3\times3}$

the calculus of variations, as in Bryson and Ho [35], or Kwakernaak and Sivan [113] or by Pontryagin's maximum principle, as in Athans and Falb [12]. The calculus-of-variations approach leads directly to the linear equations of Section 2.4, while the dynamic-programming approach leads directly to the Riccati equation of Section 2.3. For more on dynamic programming and the principle of optimality see Bellman [18] and Bellman and Dreyfus [19]. For an interesting link between the Hamilton–Jacobi equations obtained from dynamic programming and the calculus of variations see Caratheodory [39]. For a survey of results and outstanding problems in linear-quadratic control see Casti [40].

Some of the references that deal with the theory of matrix Riccati equations include [36], [38], [89], [100], [131], [172], [186], and [206]; while those dealing with the theory of algebraic Riccati equations include [94], [114], [143], [169], and [205]. Many papers and some books have been written on the numerical solution of Riccati equations, including [8], [63], [90], [108], [115], [125], [142], and [167]. Riccati equations have appeared in many systems-engineering applications. See, e.g., the volume edited by Bittanti [30], on applications

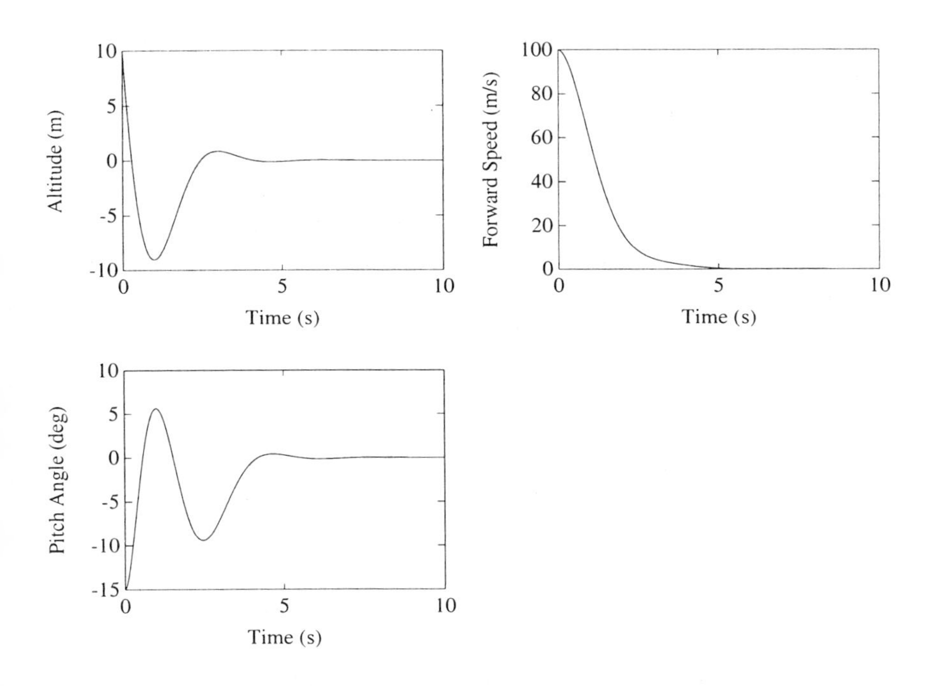

Figure 2.4: Outputs for the Aircraft, $q_{11} = 100$, $r_{11} = 100$

to control systems and signal processing.

A basic assumption we make in the development of LQR theory is that the matrix R is nonsingular. Indeed, all of the feedback-control laws developed in this chapter require the inverse of R. There are, however, some interesting control problems where this matrix might be singular. These problems are commonly referred to as **singular control problems**. See Clements and Anderson [47] or Jacobson [96] for books that deal with this special problem. See also the related works by Francis and Wonham [68] and Isermann [95]. In Molinari [144] and Willems [206], the general LQR problem is studied where Q and R do not satisfy the usual conditions $Q \geq 0$ and $R > 0$.

Many papers have also been written on the selection of the matrices Q, R, and M. Obviously, every application paper must deal with this problem. We cite here just a few references on the subject, in particular [66], [85], [138], [189], and [197]. It is interesting to note that the concept of cheap control has a number of very important applications, e.g., to robust and stochastic control problems. Francis [66] and Saberi and Sannuti [177] deal further with the problem of cheap control, including the case where the plant may not be minimum phase. It is also possible to introduce **frequency-shaped** R and Q matrices in order to control the frequency response of the optimal feedback system. For an example of frequency-shaping designs, see, e.g., chap. 9 of Anderson and Moore [4].

We have assumed in all of our discussions on the LQR problem that the system dynamics are given in the standard state-space form (equation (2.2)). For certain problems, however, e.g., singularly perturbed systems, nonproper plants, etc., the more general **descriptor** (also known as singular or generalized) form

$$E\dot{x} = Ax + Bu$$

is required, where the matrix E is singular. A discussion of optimal LQR design for this class of plants may be found in Bender and Laub [20].

We have also assumed that the state of the plant is available for feedback control. In Levine, Johnson, and Athans [120], gradient methods are used to solve LQR problems where only the output of the plant is available for feedback. The use of gradient methods to solve constrained feedback problems of this type is discussed in Sections 6.3 and 6.5.

Finally, it should be noted that many applications of LQR design have been reported in conference papers that are not referenced here. Many applications to aerospace problems may be found in the text of McLean [137]. Some applications of LQR design for process control may be found in the texts of Balchen and Mumme [14] and Ray [171].

2.10 Problems

Problem 2.1 Use the eigenvalue–eigenvector approach of Section 2.4 to solve the following LQR problem for arbitrary α, β, and T

Minimize

$$V = \int_0^T (x^2 + \alpha u^2)dt + \beta x^2(T)$$

given the dynamics $\dot{x} = u$. In particular, compute the optimal feedback gain K and the optimal performance-measure value when $x(0) = 0.5$.

Problem 2.2 Use the solution of the algebraic Riccati equation, (2.45), to compute the feedback gain K for the LQR problem

Minimize

$$V = \int_0^\infty [(x_1)^2 + u^2]dt$$

where the state x has components x_1 and x_2 that satisfy the equations $\dot{x}_1 = x_2$ and $\dot{x}_2 = u$.

Problem 2.3 Solve the matrix linear equation (2.60) to compute the control law that drives the state of the system in problem 2.2 to zero and minimizes

$$V = \int_0^T u^2 \, dt$$

Problem 2.4 Consider the problem of minimizing

$$V = \int_t^T (x^2 + u^2)dt + \rho x^2(T)$$

given the scalar system $\dot{x} = u$. Use separation of variables to solve the associated scalar Riccati equation for arbitrary ρ. Sketch $P(t)$ for all $t \leq T$ for each of the following values of $\rho : 2, 1, 0, -1$, and -2. Note that letting t approach $-\infty$ is equivalent to letting T approach $+\infty$. Is $P(t)$ monotonically increasing with T in each case? Does this contradict the result in Section 2.5 about the monotonicity of $P(t)$? Explain. Does a steady-state solution exist in each case? Explain.

Problem 2.5 Using the eigenvalue–eigenvector method, find the feedback matrix K for a steady-state LQR problem with the following data:

$$A = \begin{bmatrix} 0 & 10 \\ -10 & 0 \end{bmatrix}; \; B = Q = R = I$$

where I is the 2×2 identity matrix. Use the **eig** function in MATLAB if necessary.

Problem 2.6 Redesign problem 2.2 for a degree of stability equal to minus 10.

Problem 2.7 Consider further the undetectable example in Section 2.5, i.e.,

$$V = \int_0^\infty u^2 \, dt, \quad \dot{x} = x + u$$

The ARE for this problem has two solutions, $P = 0$ and $P = 2$, with corresponding control laws $u = 0$ and $u = -2x$. Show that both solutions appear to satisfy the sufficient condition (2.47). Obviously, only the solution $P = 0$ can be minimal. How do you explain this? Start with the stabilizing control law $u = -3x$, and do three iterations of the approximation-in-policy-space algorithm. What control law is being approached by these iterations?

Problem 2.8 Derive the optimal LQR solution when there is a cross-coupling term in x and u in the performance measure, i.e., $l(x, u, t) = x'Qx + 2u'Nx + u'Ru$. In particular, show that $u^* = -R^{-1}(N + B'P)x$, where P satisfies the differential equation

$$\begin{aligned} -\dot{P} &= (A - BR^{-1}N)'P + P(A - BR^{-1}N) \\ &\quad + (Q - N'R^{-1}N) - PBR^{-1}B'P \end{aligned}$$

Note that for P to have all the usual properties, i.e., be positive definite, stabilizing, etc., we now require in addition to $R > 0$ that $Q - N'R^{-1}N > 0$.

Problem 2.9 Consider a system describing an automobile gas turbine [91] with the following matrices:

$$A = \begin{bmatrix} A_1 & 0 & 0 & 0 \\ 0 & A_2 & 0 & 0 \\ 0 & 0 & A_3 & 0 \\ 0 & 0 & 0 & A_4 \end{bmatrix}$$

where

$$A_1 = \begin{bmatrix} 0 & 1 \\ -0.202 & -1.15 \end{bmatrix}; \; A_2 = \begin{bmatrix} 0 & 1 & 0 \\ 0 & 0 & 1 \\ -2.36 & -13.6 & -12.8 \end{bmatrix}$$

$$A_3 = \begin{bmatrix} 0 & 1 & 0 \\ 0 & 0 & 1 \\ -1.62 & -9.4 & -9.15 \end{bmatrix}$$

$$A_4 = \begin{bmatrix} 0 & 1 & 0 & 0 \\ 0 & 0 & 1 & 0 \\ 0 & 0 & 0 & 1 \\ -188 & -111.6 & -116.4 & -20.8 \end{bmatrix}$$

$$B = \begin{bmatrix} 0 & 1.0439 & 0 & 0 & -1.794 & 0 & 0 & 1.0439 & 0 & 0 & 0 & -1.794 \\ 0 & 4.1486 & 0 & 0 & 2.6775 & 0 & 0 & 4.1486 & 0 & 0 & 0 & 2.6775 \end{bmatrix}'$$

$$C = [C_1 \; C_2]$$

where

$$\begin{aligned} C_1 &= \begin{bmatrix} 0.264 & 0.806 & -1.42 & -15 & 0 & 0 \\ 0 & 0 & 0 & 0 & 0 & 4.9 \end{bmatrix} \\ C_2 &= \begin{bmatrix} 0 & 0 & 0 & 0 & 0 & 0 \\ 2.12 & 1.95 & 9.35 & 25.8 & 7.14 & 0 \end{bmatrix} \\ D &= \begin{bmatrix} 0 & 0 \\ 0 & 0 \end{bmatrix} \end{aligned}$$

1. Is the system controllable? Is it observable? You may use the MATLAB functions **ctrb** and **obsv**.

2. As a matter of fact, this system is neither controllable nor observable. Find a minimal realization of the system using the MATLAB function **minreal**. Now compare the eigenvalues of the original system with those of the minimal one and determine whether the original system is stabilizable and/or detectable.

3. Choose $Q = C'C$ and $R = B'B$ and design a steady-state LQR controller. Compare the degree of stability of the optimal feedback system with that of the original system.

Problem 2.10 Given the following state-space system known as the double integrator

$$\dot{x} = \begin{bmatrix} 0 & 1 \\ 0 & 0 \end{bmatrix} x + \begin{bmatrix} 0 \\ 1 \end{bmatrix} u$$

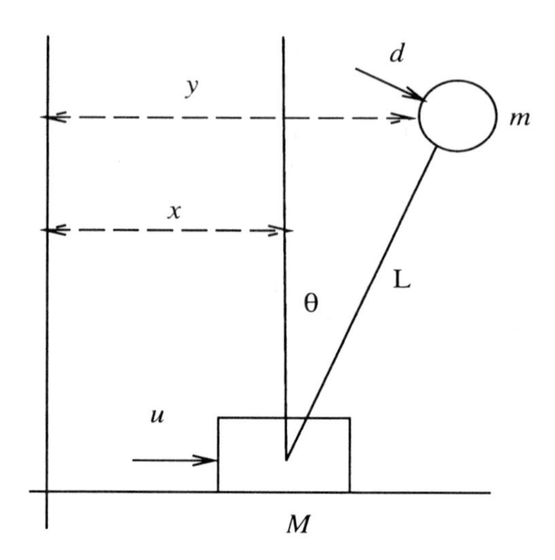

Figure 2.5: The Inverted Pendulum

Let $M =)_{2\times 2}$, $R = 1$ and $Q = I_{2\times 2}$. Show that for T finite

$$\begin{aligned} K &= [K_1 \ K_2] \\ K_1 &= \frac{cosh[\sqrt{3}(t-T)] - 1}{2 + cosh[\sqrt{3}(t-T)]} \\ K_2 &= \frac{-\sqrt{3}sinh[\sqrt{3}(t-T)]}{2 + cosh[\sqrt{3}(t-T)]} \end{aligned}$$

and that the steady-state gain is $K = [1 \ \sqrt{3}]$.

Problem 2.11 The inverted-pendulum cart shown in Figure 2.5 is linearized about the straight-up position and has the following equations when the disturbance d is perpendicular to the pendulum [57]:

$$\begin{bmatrix} \dot{x} \\ \ddot{x} \\ \dot{\theta} \\ \ddot{\theta} \end{bmatrix} = \begin{bmatrix} 0 & 1 & 0 & 0 \\ 0 & 0 & -\frac{m}{M}g & 0 \\ 0 & 0 & 0 & 1 \\ 0 & 0 & \frac{M+m}{ML}g & 0 \end{bmatrix} \begin{bmatrix} x \\ \dot{x} \\ \theta \\ \dot{\theta} \end{bmatrix} + \begin{bmatrix} 0 \\ \frac{1}{M} \\ 0 \\ -\frac{1}{ML} \end{bmatrix} u + \begin{bmatrix} 0 \\ -\frac{1}{M} \\ 0 \\ \frac{M+m}{mML} \end{bmatrix} d$$

Assume the following numerical values,

- $d = 0$
- $M = 2$ kilograms (kg)

- $m = 1$ kilograms (kg)
- $L = 0.5$ meters (m)
- $g = 9.81$ meters per second squared (m/s/s)

1. Find the steady-state LQR controller when $Q = \text{diag}(1, 1, 10, 10)$ and $R = [1]$. Repeat for $R = [10]$. Plot $\theta(t)$ and $u(t)$ for both cases when $x(0) = [0\ \ 0\ \ 1\ \ 0]'$.
2. Design for $R = [1]$ the steady-state LQR controller to achieve a degree of stability given by $\alpha = 5$. Compute the closed-loop eigenvalues to verify your design.

Problem 2.12 The following equations describe a fifth-order model of a drum boiler [21] with a maximum steam flow of about 350 t/h, a drum pressure of 140 bar and an operating point of 90 percent full load:

$$A = \begin{bmatrix} -0.129 & 0 & 0.0396 & 0.025 & 0.0191 \\ 0.0329 & 0 & -0.000078 & -0.000122 & -0.621 \\ 0.0718 & 0 & -0.1 & 0.000887 & -3.851 \\ 0.0411 & 0 & 0 & -0.0822 & 0 \\ 0.000361 & 0 & 0.000035 & 0.0000426 & -0.0743 \end{bmatrix}$$

$$B = \begin{bmatrix} 0 & 0.00139 \\ 0 & 0.0000359 \\ 0 & -0.00989 \\ 0.0000249 & 0 \\ 0 & -0.00000543 \end{bmatrix} ; \ C = \begin{bmatrix} 1 & 0 & 0 & 0 & 0 \\ 0 & 1 & 0 & 0 & 0 \end{bmatrix}$$

where

- x_1 is the drum pressure (bar)
- x_2 is the drum liquid level (m)
- x_3 is the drum liquid temperature (degrees Celcius)
- x_4 is the riser wall temperature (degrees Celcius)
- x_5 is the steam quality (percent)
- u_1 is the heat flow to the risers (kJ/s)

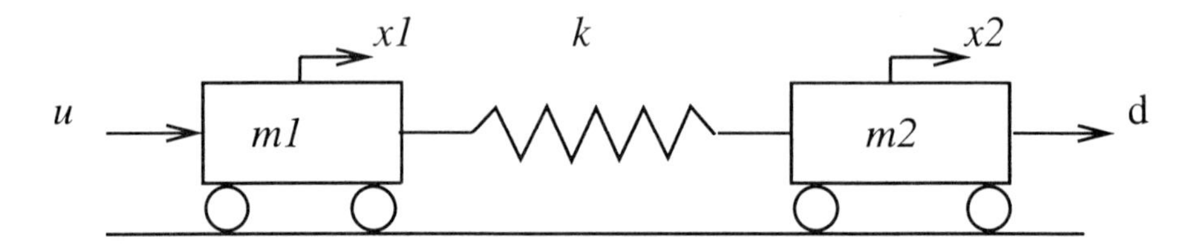

Figure 2.6: Two-Mass/Spring System

- u_2 is the feedwater flow (kg/s)

Find the steady-state optimal LQR state-feedback gain K when $Q = I_{5\times5}$ and $R = 10I_{2\times2}$.

Problem 2.13 Given the two-mass spring system shown in Figure 2.6 and described by the following:

$$\begin{bmatrix} \dot{x}_1 \\ \dot{x}_2 \\ \dot{x}_3 \\ \dot{x}_4 \end{bmatrix} = \begin{bmatrix} 0 & 0 & 1 & 0 \\ 0 & 0 & 0 & 1 \\ \frac{-k}{m_1} & \frac{k}{m_1} & 0 & 0 \\ \frac{k}{m_2} & \frac{-k}{m_2} & 0 & 0 \end{bmatrix} \begin{bmatrix} x_1 \\ x_2 \\ x_3 \\ x_4 \end{bmatrix} + \begin{bmatrix} 0 \\ 0 \\ \frac{1}{m_1} \\ 0 \end{bmatrix} u + \begin{bmatrix} 0 \\ 0 \\ 0 \\ \frac{1}{m_2} \end{bmatrix} d$$

where

- x_1 is the position of body 1 (m)
- x_2 is the position of body 2 (m)
- x_3 is the velocity of body 1 (m/s)
- x_4 is the velocity of body 2 (m/s)
- u is the control-force input (N)
- d is a disturbance-force input (N)

This system was introduced in [202] as a case study for different robust controllers. Assume the following nominal values for the different parameters of the system: $m_1 = m_2 = k = 1$, and $d = 0$.

1. Design a steady-state LQR controller for $Q_1 = \text{diag}(1, 0, 0, 0)$ and $R = [1]$.
2. Compare the transient behavior of the state components $x_1(t)$ and $x_2(t)$ and the control effort $u(t)$, given the initial disturbance $x(0) = [1 \;\; 0 \;\; 0 \;\; 0]'$ and the control law above, with that obtained if Q_1 were changed to $Q_2 = \text{diag}(10, 0, 0, 0)$.

Chapter 3

Tracking & Disturbance Rejection

In this chapter we study problems where the ideal state is not the zero state (tracking problems) or where a disturbance signal is present in the system dynamics (disturbance-rejection problems). It is shown that the optimal linear-quadratic control law in this case involves a feedforward term in addition to the usual state-feedback term. In addition, we briefly discuss the general unknown disturbance case, and end with the proportional-integral approach to rejecting an unknown but constant disturbance.

3.1 Tracking Problem

The tracking problem is by definition the problem of minimizing the performance measure

$$V = \int_t^T [(\xi - \tilde{\xi})'Q(\xi - \tilde{\xi}) + u'Ru]d\tau \tag{3.1}$$

where the desired state is $\tilde{\xi}$ and the system state ξ is given by the usual linear dynamics

$$\dot{\xi} = A\xi + Bu \tag{3.2}$$

In the above problem we assume that the desired state $\tilde{\xi}$ is known for all $\tau \geq t$ and that the system state ξ is available for feedback. Also, for convenience we have omitted a terminal-error term in (3.1). This term can easily be added as shown later.

Some tracking problems can be reduced to regulator problems, studied in the previous chapter. For example, if the desired state and the system matrix A are such that

$$A\tilde{\xi} - \dot{\tilde{\xi}} = 0 \tag{3.3}$$

then the substitution $x = \xi - \tilde{\xi}$ yields the system dynamics

$$\dot{x} = Ax + Bu + (A\tilde{\xi} - \dot{\tilde{\xi}}) = Ax + Bu \tag{3.4}$$

and the performance measure

$$V = \int_t^T (x'Qx + u'Ru)d\tau \tag{3.5}$$

which is precisely the regulator problem defined in Chapter 2.

The optimal feedback control law in this case is simply

$$u^*(t) = -K(t)(\xi(t) - \tilde{\xi}(t)) \tag{3.6}$$

where K(t) is the solution of the regulator problem as given by (2.23) and (2.24). Equation (3.3) is an example of the **internal model principle** [68], which basically states that for proper asymptotic tracking the plant must model the non asymptotically stable modes of the reference signal (desired state trajectory). Note, for example, that if

$$\tilde{\xi} = e^{\lambda t}v$$

where λ is an eigenvalue of A with corresponding eigenvector v, then equation (3.3) is satisfied even if λ has a positive real part. Tracking problems that can be reduced to regulator problems of this type are sometimes called **servomechanism problems**.

As will be shown in the next section, tracking problems that cannot be reduced to regulator problems can always be reduced to an equivalent disturbance-rejection problem. The disturbance-rejection problem is formulated and solved in the next two sections.

3.2 Disturbance-Rejection Problem

In the disturbance-rejection problem it is assumed that the system dynamics $\dot{x} = Ax + Bu$ are additively disturbed by a signal $w(t)$, i.e., that the system dynamics are given by

$$\dot{x} = Ax + Bu + w(t) \tag{3.7}$$

and that the objective is to determine the control input that minimizes the effect of this disturbance on the value of the performance index

$$V = \int_t^T (x'Qx + u'Ru)d\tau \tag{3.8}$$

We assume for the moment that the disturbance signal is known for all $\tau \geq t$. Note that if in the tracking problem

$$A\tilde{\xi} - \dot{\tilde{\xi}} = w(t) \neq 0 \tag{3.9}$$

then the tracking problem in the variable $x = \xi - \tilde{\xi}$ reduces to the disturbance-rejection problem defined above. We illustrate this reduction next by a simple example.

Example 3.1 Consider a tracking problem for the double-integrator system with dynamics

$$\dot{\xi} = \begin{bmatrix} 0 & 1 \\ 0 & 0 \end{bmatrix} \xi + \begin{bmatrix} 0 \\ 1 \end{bmatrix} u$$

and a desired state trajectory

$$\tilde{\xi} = \begin{bmatrix} 1 \\ t \end{bmatrix}$$

In this case the tracking problem reduces to a disturbance-rejection problem with a disturbance signal

$$w(t) = A\tilde{\xi} - \dot{\tilde{\xi}} = \begin{bmatrix} t \\ -1 \end{bmatrix}$$

Note that if the desired state were instead

$$\tilde{\xi} = \begin{bmatrix} t \\ 1 \end{bmatrix}$$

then the tracking problem would reduce to a regulator problem since now $w(t) = 0$.

3.3 Disturbance-Rejection Solution

We consider now the problem of solving the disturbance-rejection problem represented by the dynamics given in (3.7) and the performance measure given in (3.8). Recall that this also solves the tracking problem when $w(t)$ is given as in (3.9), and $x = \xi - \tilde{\xi}$. A simple quadratic form such as $x'Px$ for the solution of the Hamilton–Jacobi equation will not work in this case because of the disturbance term $w(t)$ in the system dynamics. A more general quadratic form is required, in particular,

$$V^*(x,t) = x'P(t)x + 2b'(t)x + c(t) \tag{3.10}$$

The Hamilton–Jacobi equation becomes in this case

$$-x'\dot{P}x - 2\dot{b}'x - \dot{c} = \min_u \left\{ x'Qx + u'Ru + \left[\frac{\partial V^*}{\partial x}\right]' (Ax + Bu + w) \right\} \tag{3.11}$$

with $\partial V^*/\partial x = 2(Px + b)$. If the minimum of the right-hand term in (3.11) is computed in the usual way by setting the gradient with respect to u equal to zero, one obtains

$$u^* = -R^{-1}B'[P(t)x + b(t)] \tag{3.12}$$

If next we substitute u^* given by (3.12) into (3.11) and equate different "powers" of x we obtain, after some matrix manipulations

Quadratic terms in x

$$-\dot{P} = A'P + PA + Q - PBR^{-1}B'P \tag{3.13}$$

Linear terms in x

$$\dot{b} = -[A - BR^{-1}B'P(t)]'b - P(t)w(t) \tag{3.14}$$

Terms independent of x

$$\dot{c} = b'(t)BR^{-1}B'b(t) - 2b'(t)w(t) \tag{3.15}$$

Finally, from the boundary condition on V^*, i.e.,

$$V^*(x,T) = 0, \text{ for all } x$$

we obtain the following boundary conditions for each of the above differential equations

$$P(T) = 0; \ b(T) = 0; \ c(T) = 0 \tag{3.16}$$

If $M \neq 0$, the only change required in the boundary conditions of (3.16) is that $P(T) = M$. The solution of (3.13), (3.14), and (3.15) subject to the boundary conditions given in (3.16) constitutes the solution of the disturbance-rejection problem (and, by reduction, the state tracking problem). Fortunately, the above differential equations for $P(t), b(t)$, and $c(t)$ can be solved sequentially, i.e., one can solve (3.13) for $P(t)$, then use this solution in (3.14) to solve for $b(t)$, and then use this solution in (3.15) to solve for $c(t)$. Unfortunately, it is difficult to go analytically beyond the solution for $P(t)$ since even when all the data is time-invariant, the differential equation for $b(t)$ is time-varying, and general analytic solutions are not available for time-varying differential equations. At best one can express the solution of $b(t)$ in terms of the transition matrix for the system

$$\dot{b} = -[A - BR^{-1}B'P(t)]'b \tag{3.17}$$

which must generally be obtained numerically. Let the transition matrix for the time-varying differential equation (3.17) be denoted $\Phi_b(t, \tau)$, then the solution of (3.14), subject to the boundary condition in (3.16), can be written

$$b(t) = \int_t^T \Phi_b(t, \tau)P(\tau)w(\tau) \, d\tau \tag{3.18}$$

It is clear from (3.18) that the solution of the disturbance-rejection problem as stated above requires knowledge of the disturbance input over the entire optimization interval (t, T). Thus, the tracking problem also requires knowledge of the desired state over the entire optimization interval. In any case, once $P(t)$ and $b(t)$ have been computed, the control law may be written,

$$u^*(t) = -K(t)x(t) + u_{fw}(t) \tag{3.19}$$

where the feedback matrix K is given by

$$K(t) = R^{-1}B'P(t) \tag{3.20}$$

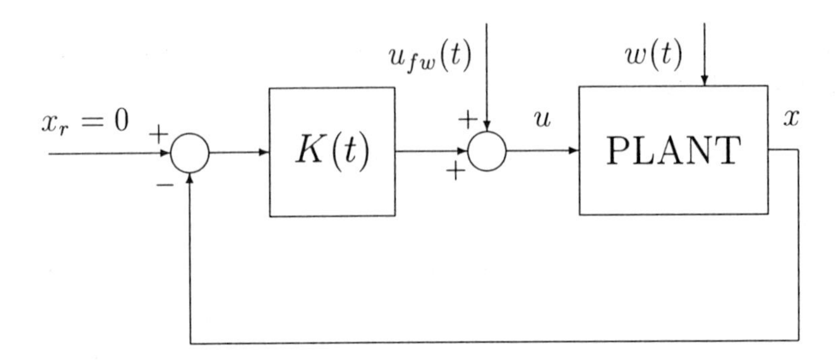

Figure 3.1: Optimal LQR Disturbance-Rejection Configuration

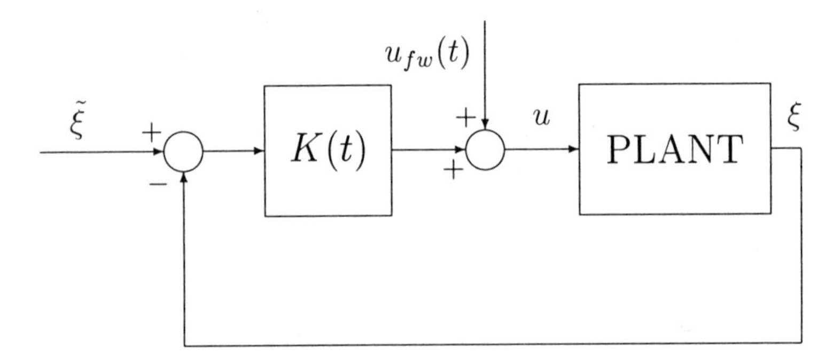

Figure 3.2: Optimal LQR Tracking Configuration

and the feedforward signal $u_{fw}(t)$ is given by

$$u_{fw}(t) = -R^{-1}B'b(t) \tag{3.21}$$

Figure 3.1 shows the optimal LQR disturbance-rejection configuration, while Figure 3.2 shows the optimal tracking configuration, when the tracking problem cannot be reduced to a regulator problem.

Note that the matrix $P(t)$ satisfies the same Riccati equation as that of the regulator problem. Indeed, if the disturbance is zero, the disturbance-rejection problem reduces to the regulator problem, as expected. Finally, it should be noted that $c(t)$ plays no role in determining the control law. It contributes only to the evaluation of the optimal performance measure. In particular, it is the value of V^* when the initial state $x(t) = x$ is zero. Of course, if the disturbance is zero, this value is also zero, as in the regulator case. Equation (3.15) may be integrated, with the constant of integration given by the boundary condition in (3.16), to obtain

$$c(t) = \int_t^T [2b'(\tau)w(\tau) - b'(\tau)BR^{-1}B'b(\tau)]d\tau \tag{3.22}$$

We return now to the problem that the feedforward term $u_{fw}(t)$ is noncausal, i.e., it depends on future values of the disturbance signal w. Of course, if $w(t)$ is indeed known for all time, the signal u_{fw} can be computed offline. If additional assumptions can be made on how the disturbance is generated, however, an online solution is possible. In particular, if $w(t)$ is assumed to be generated by a differential equation, possibly time-varying, of the form

$$\dot{w} = A_w w \tag{3.23}$$

then u_{fw} becomes a causal function of w. In fact, we have from (3.18)

$$b(t) = \left[\int_t^T \Phi_b(t,\tau)P(\tau)\Phi_w(\tau,t)d\tau\right] w(t) \tag{3.24}$$

where $\Phi_w(\tau, t)$ is the transition matrix for (3.23). In some applications, and in order to model complex signals, the state of the disturbance signal may have to be of higher order than the state of the system. In that case, (3.23) may be replaced by

$$\dot{\psi} = A_\psi \psi, \; w(t) = C_\psi \psi(t) \tag{3.25}$$

and $b(t)$ by

$$b(t) = \left[\int_t^T \Phi_b(t,\tau)P(\tau)C_\psi\Phi_\psi(\tau,t)d\tau\right] \psi(t) \tag{3.26}$$

In this case $b(t)$ is a causal function of the "augmented" disturbance state $\psi(t)$. It is assumed that this state can be measured.

We consider next a simple disturbance-rejection example to illustrate some of the difficulties involved in solving tracking and disturbance rejection problems.

Example 3.2 Consider the problem of minimizing

$$V = \int_0^T (x^2 + u^2)dt$$

subject to

$$\dot{x} = u + w(t)$$

where the disturbance w is assumed to satisfy the differential equation $\dot{w} = w$.

Solution: From Example 2.1, the solution of $P(t)$ for the above problem is

$$P(t) = \frac{1 - e^{-2(T-t)}}{1 + e^{-2(T-t)}}$$

and the transition matrix for equation $\dot{w} = w$ is

$$\Phi_w(\tau, t) = e^{(\tau - t)}$$

while the transition matrix for b is

$$\Phi_b(t, \tau) = e^{-\int_t^\tau P(\alpha) d\alpha}$$

Now while all the functions needed to compute $b(t)$, from equation (3.24), are theoretically available, the integration cannot be done analytically, even for this simple scalar example.

3.4 Steady-State and Preview Control

We consider now two special cases for the final time T. One is for $T = \infty$, **steady-state control**, and the other is for $T = t + T_p$, **preview control**, also sometimes referred to as **receding-horizon control**. The concept of steady-state control has already been discussed in terms of the regulator problem in the last chapter. The basic idea of preview control is that the final time in the optimization problem is actually a **moving time** with a fixed time-to-go equal to T_p. This choice of "floating" T is a bit strange, since one only starts to optimize for T_p units of time into the future at any given time t, but one never actually operates optimally over any period of time. However, this problem has been found to be useful in some applications, especially where the desired state or disturbance signal at time t is known for only a finite interval T_p. We discuss first the steady-state problem.

Steady-State Control. The data matrices A, B, A_w, Q, and R are assumed time-invariant. It is further assumed that all the conditions for a steady-state solution of the Riccati equation (3.13), as discussed in Chapter 2, are satisfied. Let this steady-state solution be denoted $\bar{P}$, and let $A_c = A - BR^{-1}B'\bar{P}$, as in Section 2.5. Then $b(t)$ can be computed from (3.24), for $T = \infty$, i.e.,

$$b(t) = K_{fw}w(t) \tag{3.27}$$

where K_{fw} is the constant matrix given by

$$K_{fw} = \int_0^\infty e^{A_c'\alpha}\bar{P}e^{A_w\alpha}d\alpha \tag{3.28}$$

Note that K_{fw} given by (3.28) can be computed analytically, since the terms in the integrand are matrix exponentials; however, K_{fw} is finite only if the eigenvalues of A_c all have real parts that are sufficiently negative to overcome all the eigenvalues of A_w with positive real parts. We present, however, an algebraic approach to computing K_{fw} when it is finite, using a nonstandard Lyapunov equation.

The gain K_{fw} of (3.28) may be found as the solution of

$$0 = A_c'K_{fw} + K_{fw}A_w + \bar{P} \tag{3.29}$$

assuming that

$$\lim_{\alpha\to\infty} e^{A_c'\alpha}\bar{P}e^{A_w\alpha} = 0$$

Proof: Let us premultiply (3.29) by $e^{A_c'\alpha}$ and postmultiply it by $e^{A_w\alpha}$ to obtain

$$0 = e^{A_c'\alpha}A_c'K_{fw}e^{A_w\alpha} + e^{A_c'\alpha}K_{fw}A_we^{A_w\alpha} + e^{A_c'\alpha}\bar{P}e^{A_w\alpha} \tag{3.30}$$

Now noting that $Ge^{G\alpha} = e^{G\alpha}G$ for any square G, we write

$$\frac{d}{d\alpha}\left(e^{A_c'\alpha}K_{fw}e^{A_w\alpha}\right) = e^{A_c'\alpha}A_c'K_{fw}e^{A_w\alpha} + e^{A_c'\alpha}K_{fw}A_we^{A_w\alpha}$$

Then, (3.30) becomes

$$0 = \frac{d}{d\alpha}\left(e^{A_c'\alpha}K_{fw}e^{A_w\alpha}\right) + e^{A_c'\alpha}\bar{P}e^{A_w\alpha}$$

which we integrate to obtain

$$\begin{aligned} 0 &= \int_0^\infty \left[\frac{d}{d\alpha} \left(e^{A_c' \alpha} K_{fw} e^{A_w \alpha} \right) + e^{A_c' \alpha} \bar{P} e^{A_w \alpha} \right] d\alpha \\ &= \left[e^{A_c' \alpha} K_{fw} e^{A_w \alpha} \right]_0^\infty + \int_0^\infty e^{A_c' \alpha} \bar{P} e^{A_w \alpha} d\alpha \\ &= -K_{fw} + \int_0^\infty e^{A_c' \alpha} \bar{P} e^{A_w \alpha} d\alpha \end{aligned}$$

which is exactly (3.28). ■

Note also that $b(t)$ given by (3.27) is not bounded when $w(t)$ is unstable. Finally, note that $c(t)$ is not likely to be bounded as T approaches ∞. To see this, consider the simple case where $w(t) = w_0$. This is equivalent to $\Phi_w(t, \tau) = I$. In this case, and as T goes to ∞, $c(t)$ approaches

$$c(t) = (T - t) w_0' (K_{fw}' + K_{fw} - K_{fw}' B R^{-1} B' K_{fw}) w_0 \qquad (3.31)$$

which obviously becomes unbounded as T goes to ∞. In fact, for most disturbance-rejection and tracking problems the performance measure must be modified if it is to remain finite. Typically, one divides V by T to keep V^* finite.

Preview Control. If we let $T = t + T_p$ in (2.40), we see that $P(t)$ is independent of t. This means that the feedback matrix is time-invariant, even for optimization problems over a finite time interval. This is one of the reasons for using preview control. The value of $P(t)$ is then given by the solution $\bar{P}$ of the ARE, e.g., equation (2.45). Note, however, that to satisfy the boundary condition $P(T) = M$, we must have $M = \bar{P}$. This is an artificial constraint on the choice of performance measure but is required for a constant solution of the Riccati equation. Finally, it should be noted that with the usual assumptions on the plant and performance measure, the closed-loop system is stable, since the solution $\bar{P}$ of (2.45) results in a matrix $A - BR^{-1}B'\bar{P}$, which has all eigenvalues in the left-half plane.

This finite previewing can be realized in a practical way in many applications, for example, with "look-ahead" radar in flight-control problems, with "feel-ahead" sensors in robotic problems, etc. Of course, as previously noted, a disadvantage of such control is that it is never truly optimal over any fixed interval of time. We consider next a simple example to illustrate preview control.

Example 3.3 Consider a preview problem with

$$V = \int_t^{t+T_p} (x^2 + u^2)d\tau + Mx^2(T)$$

subject to

$$\dot{x} = u + w(t)$$

Solution: The ARE in this case is $0 = 1 - P^2$, which has the positive solution $\bar{P} = 1$; thus, $P(t) = \bar{P} = 1$ must be the constant solution of the preview problem, and of necessity the terminal weight M in the above performance measure must be $M = 1$. Furthermore, $\Phi(t, \tau) = e^{t-\tau}$, and $b(t)$ given by (3.18) may be written

$$b(t) = \int_t^{t+T_p} e^{(t-\tau)} w(\tau)\, d\tau$$

The final optimal control law is then given by

$$u^*(t) = -Kx(t) + u_{fw}(t) = -x(t) - b(t)$$

where $b(t)$ is given by the above integral.

3.5 Unknown Disturbances

In all of the previous sections we have assumed that the disturbance signal or desired state was known over the entire optimization interval. In many practical applications this is not the case. Although we do not go into details here, we mention some approaches to the problem when these signals are not known.

Stochastic Approach. One approach is to model the uncertain disturbances as stochastic processes and to measure performance by taking the expected value of the usual integral-quadratic performance measure. The problem then becomes a stochastic control problem. Such problems are discussed in more detail in Chapter 5.

State-Estimation Approach. In some problems it is possible to design an"observer" that can estimate the state of the uncertain disturbance. In this case the feedforward gain typically becomes a dynamical system. State estimation is discussed in Chapter 6.

Minimax Approach. It is also possible to do a "worst-case" design and assume the disturbance signal is **maximizing** the same performance measure that the control input is **minimizing**. If we assume that the disturbance signal has the same information as the control signal, such problems reduce to optimal game-theoretic problems that can be solved using the same kind of dynamic-programming arguments used in solving the LQR problem in Chapter 2. A typical formulation of a problem of this type is to solve the following **differential-game problem**. Find the control u and disturbance w such that

$$V^* = \min_u \max_w \int_t^T (x'Qx + u'Ru - w'Ww)dt \tag{3.32}$$

where x satisfies the equation

$$\dot{x} = Ax + Bu + Gw \tag{3.33}$$

The matrix G is added in the dynamics (3.33) to generalize the problem a bit, and the term $-w'Ww$ is included in the performance measure (3.32) as a Lagrangian multiplier term to solve a problem of minimaxing

$$V = \int_0^T (x'Qx + u'Ru)dt$$

subject to the constraint

$$\int_0^T w'Ww \, dt = 1$$

Thus, the only information assumed about the disturbance is that its "integral-squared" energy is bounded. We give here only a very terse account of the solution of such problems. A more detailed discussion may be found in Bryson and Ho [35]. First, it can be shown that under appropriate conditions, the principle of optimality applies to the minimaxing operation in (3.32) and that one has for this problem the following **game-theoretic Hamilton–Jacobi optimization equation**

$$
\begin{aligned}
-\frac{\partial V^*}{\partial t} &= \min_u \max_w \left\{ x'Qx + u'Ru - w'Ww \right. \\
&\quad \left. + \left[\frac{\partial V^*}{\partial x}\right]'(Ax + Bu + Gw) \right\}
\end{aligned} \tag{3.34}
$$

By taking gradients of the square-bracketed term in (3.34) with respect to u and w one obtains

$$
u^* = -\frac{1}{2}R^{-1}B'\frac{\partial V^*}{\partial x}; \quad w^* = \frac{1}{2}W^{-1}G'\frac{\partial V^*}{\partial x} \tag{3.35}
$$

If we let $V^*(x,t) = x'P(t)x$ and substitute the values of u^* and w^* given in (3.35) back into (3.34), we obtain the "nonstandard" Riccati differential equation

$$
\begin{aligned}
-\dot{P} &= A'P + PA + Q - P(BR^{-1}B' - GW^{-1}G')P, \\
P(T) &= 0
\end{aligned} \tag{3.36}
$$

We label this Riccati equation nonstandard because the "standard" positive-semidefinite term $BR^{-1}B'$ has been replaced by a term that may be indefinite or negative-semidefinite. This Riccati equation has many interesting properties that we will not pursue here. It is worth noting, however, that an equation of this type appears in the state-feedback H^∞ optimization problem. Indeed, the optimal H^∞ problem may be viewed as a disturbance-rejection problem with disturbances that are quadratically bounded (see, for example, [162]).

Proportional-plus-Integral (PI) Control. It is well known from classical control theory that integration in the loop can be used to asymptotically reject constant disturbances. We show next how LQR theory can be used to deal with problems where the unknown disturbance $w(t)$ is constant in (3.7), say $\bar{w}$. We first note that if the control is to adjust to a constant disturbance in the steady state, it cannot go to zero as it normally does in the LQR problem. The same is true for the state. It therefore makes sense to define new variables for LQR optimization. What we would like is that $\dot{u}$ and $\dot{x}$ go to zero in such a way that $Dx(t)$ goes to zero and $u(t)$ approaches a constant value that automatically adjusts for the disturbance $\bar{w}$. We assume as usual that Q has been factored into the product $D'D$. *We can now show that if the matrix $DA^{-1}B$ (assumed square for* *simplicity) is nonsingular, the following LQR optimization problem will solve our constant-disturbance-rejection problem.* Consider the problem of minimizing

$$V = \int_0^\infty [(Dx)'(Dx) + v'R_v v]dt \tag{3.37}$$

subject to the state dynamics

$$\dot{z} = \begin{bmatrix} A & 0 \\ D & 0 \end{bmatrix} z + \begin{bmatrix} B \\ 0 \end{bmatrix} v \tag{3.38}$$

where the new state z has as its first component $\dot{x}$ and its second component Dx, and $v = \dot{u}$. Note that equation (3.38) is independent of the constant disturbance $\bar{w}$. This is a consequence of the fact that the first "row" of (3.38) is obtained by differentiating the equation $\dot{x} = Ax + Bu + \bar{w}$. The second row is an obvious identity. Note also that in terms of the augmented state z, the performance measure (3.37) may be written as

$$V = \int_0^\infty [z'Q_z z + v'R_v v]dt$$

where

$$Q_z = \begin{bmatrix} 0 & 0 \\ 0 & I \end{bmatrix}$$

If this optimal LQR problem is solved one obtains a control law that is of the form

$$v^* = -Kz = -K_1\dot{x} - K_2 Dx \tag{3.39}$$

If (3.39) is integrated we get the final **proportional-plus-integral control law**

$$u^*(t) = -K_1 x(t) - K_2 \int_0^t Dx(\tau)\, d\tau \tag{3.40}$$

where the initial condition on the integrator is set equal to zero. There is no loss of generality in doing this since we will show next that $u(t)$ converges in any case to the appropriate value to cancel the disturbance $\bar{w}$. Let $x(t)$ and $u(t)$ converge to $\bar{x}$ and $\bar{u}$, respectively. Such convergence will always occur since, with the usual assumptions, the LQR problem will have an asymptotically stable closed-loop solution so that the state with the above components will asymptotically go to zero, i.e., $\dot{x}$ and $\dot{u}$ will approach zero; hence, x and u will approach constants. With this convergence the original state equation $\dot{x} = Ax + Bu + \bar{w}$ becomes asymptotically

$$0 = A\bar{x} + B\bar{u} + \bar{w} \tag{3.41}$$

If we now multiply (3.41) on the left by DA^{-1} (assuming A^{-1} exists as discussed later), and use the fact that $D\bar{x} = 0$, we obtain the equation

$$0 = DA^{-1}B\bar{u} + DA^{-1}\bar{w} \tag{3.42}$$

which, with the assumption that $DA^{-1}B$ is nonsingular, results in the fact that u converges to a unique value that automatically cancels the effect of $\bar{w}$. Note that the above nonsingularity condition is equivalent to the requirement that the transfer-function matrix $D(sI - A)^{-1}B$ has no zeros at the origin, a condition obviously required if integration in the loop is to be effective. Finally, it should be noted that the LQR optimization defined above for the PI problem is purely an artifice to obtain a stable PI control law that asymptotically rejects the constant disturbance $\bar{w}$. It does not minimize the effects of the disturbance on V, given by (3.37), since the constant disturbance enters in the initial state of the system (3.38), and the optimal LQR solution minimizes only for given initial conditions.

3.6 MATLAB Software

Except for PI control, no MATLAB software directly solves the type of tracking and disturbance-rejection problems considered in this chapter. However, one can use the functions **ode23** (or **ode45**, or **quad**) to numerically solve the differential equations for $b(t)$ and $c(t)$ (or the corresponding integral representations).

Also, the function **are** can be used to solve nonstandard Riccati equations required for minimax disturbance rejection. In particular,

$$X = are(A, G, H)$$

will solve an algebraic Riccati equation of the form

$$0 = A'X + XA + H - XGX$$

where G is positive-semidefinite but H is not necessarily positive semidefinite. This is not quite the form required in (3.36), where G is the arbitrary matrix, but by multiplying (3.36) on the right and left by the inverse of P and replacing A by $-A'$ we get the above nonstandard form. The final solution for P required for the minimax solution is then given by $P = X^{-1}$. We present next an example of a distillation column [171] to illustrate the PI solution to a tracking problem.

Example 3.4 Binary Distillation Column. The details of this model may be found in sect. 6.2 of [171]. It describes a multi-sidestream distillation column where the outputs are compositions of the sidestream and the inputs are draw-off rates of the sidestream. The state equations of the plant are

$$\begin{aligned} \dot{\xi} &= A\xi + Bu \\ y &= C\xi \end{aligned}$$

where

$$A = \begin{bmatrix} -0.111 & 0 & 0 & 0 & 0 & 0 \\ 0 & -0.125 & 0 & 0 & 0 & 0 \\ 0 & 0 & -0.167 & 0 & 0 & 0 \\ 0 & 0 & 0 & -0.1 & 0 & 0 \\ 0 & 0 & 0 & 0 & -0.125 & 0 \\ 0 & 0 & 0 & 0 & 0 & -0.143 \end{bmatrix}$$

$$B = \begin{bmatrix} 1 & 0 & 0 \\ 1 & 0 & 0 \\ 0 & 1 & 0 \\ 1 & 0 & 0 \\ 0 & 1 & 0 \\ 0 & 0 & 1 \end{bmatrix} ; \ C = \begin{bmatrix} 1 & 0 & 0 & 0 & 0 & 0 \\ 0 & 1 & 1 & 0 & 0 & 0 \\ 0 & 0 & 0 & 1 & 1 & 1 \end{bmatrix}$$

The outputs are

- y_1 is the overhead composition
- y_2 is the first sidestream composition
- y_3 is the second sidestream composition

and the inputs are

- u_1 is the draw-off rate of the overhead stream
- u_2 is the draw-off rate of the first sidestream
- u_3 is the draw-off rate of the second sidestream

Design a state-feedback controller that yields a stable closed-loop system and zero steady-state error to the output set point given by

$$\bar{y} = \begin{bmatrix} 0.05 \\ -0.05 \\ 0.02 \end{bmatrix}$$

Solution: Since the given set point is in the range of C, i.e., it is in the space spanned by the columns of C, one may compute a desired state $\tilde{\xi}$ corresponding to the desired set point from

$$\tilde{\xi} = C'(CC')^{-1}\bar{y}$$

Then, with $x = \xi - \tilde{\xi}$, the set-point tracking problem can be reduced to a disturbance-rejection problem with

$$\dot{x} = Ax + Bu + \bar{w}$$

where $\bar{w} = A\tilde{\xi}$. To achieve zero steady-state error, we use the PI control theory developed in Section 3.5. In our case, $D = C$, and the augmented state equations are given by

$$\dot{z} = \begin{bmatrix} A & 0_{6\times3} \\ C & 0_{3\times3} \end{bmatrix} z + \begin{bmatrix} B \\ 0_{3\times3} \end{bmatrix} v$$

with the performance measure

$$V = \int_0^\infty [z'Q_z z + v'R_v v]dt$$

with

$$Q_z = \begin{bmatrix} 0_{6\times6} & 0_{6\times3} \\ 0_{3\times6} & I_{3\times3} \end{bmatrix}$$

The optimal feedback matrix K is then partitioned into $K = [K_1 \mid K_2]$, and the final PI law is given by

$$u^*(t) = -\left[K_1 x(t) - K_2 \int_0^t Cx(\tau)d\tau\right]$$

For the choice of $R_v = I$, the following K was obtained using the function **lqr** in MATLAB,

$$
\begin{aligned}
K &= [K_1 \mid K_2] \\
K_1 &= \begin{bmatrix} 1.1794 & 0.2520 & 0.2516 & 0.1680 & 0.1679 & 0.1679 \\ -0.9226 & 1.0702 & 1.0302 & 0.2719 & 0.2717 & 0.2714 \\ -0.0878 & -0.9271 & -0.8869 & 1.1828 & 1.1583 & 1.1410 \end{bmatrix} \\
K_2 &= \begin{bmatrix} 0.8716 & 0.3837 & 0.3049 \\ -0.4827 & 0.7799 & 0.3984 \\ -0.0850 & -0.4944 & 0.8651 \end{bmatrix}
\end{aligned}
$$

Sketches of overhead and sidestream compositions versus time are shown in Figure 3.3, assuming $\tilde{\xi}(0) = 0$ so that the initial vector in our simulation is given by

$$z(0) = \begin{bmatrix} BK_1\tilde{\xi}(0) \\ \bar{y}(0) \end{bmatrix}$$

where

$$
\begin{aligned}
\tilde{\xi} &= \begin{bmatrix} 0.0500 & -0.0250 & -0.0250 & 0.0067 & 0.0067 & 0.0067 \end{bmatrix}' \\
\bar{y}(0) &= \begin{bmatrix} -0.05 & 0.05 & -0.02 \end{bmatrix}'
\end{aligned}
$$

3.7 Notes and References

More details on tracking and disturbance-rejection problems may be found in chap. 4 of Anderson and Moore [4]. In particular, the problem of **output** tracking, rather than **state** tracking considered here, is considered in some detail in [4]. As previously noted, when unknown disturbance signals are modeled as stochastic processes, stochastic control theory may be used to solve LQR disturbance-rejection problems. We consider the stochastic control problem in Chapter 5. The links between game theory, H^∞ control, and worst-case LQR (or equivalently H^2) control are discussed briefly in Doyle et al. [58]. The differential-game approach to LQR problems is discussed in more detail in the text of Bryson and Ho [35]. A general development of the differential-game approach to control problems may be found in the monograph by Başar and Bernhard [17].

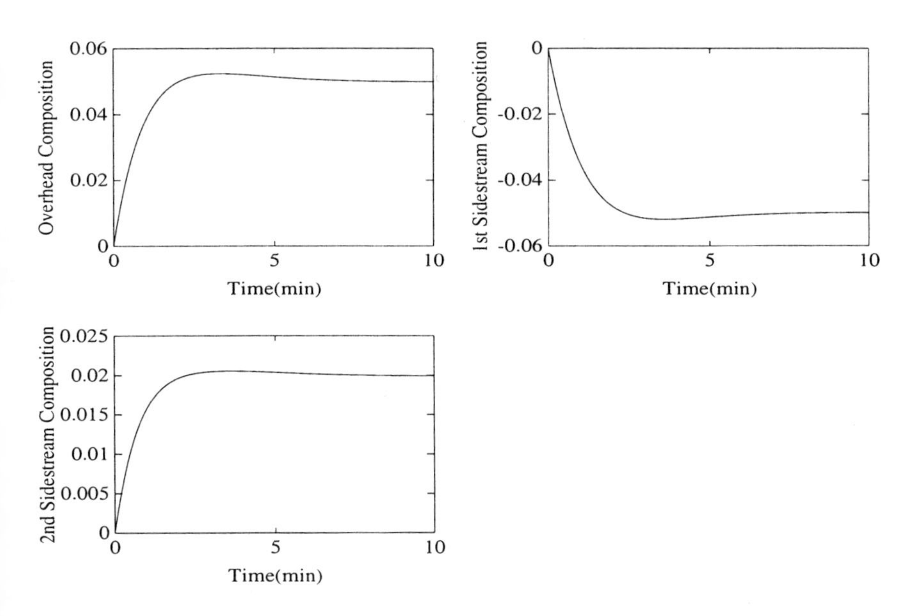

Figure 3.3: Closed-Loop Outputs for the Binary Distillation Column

Finally, it should be noted that a general theory of **robust servomechanism** design for the tracking problem in the presence of unknown disturbances is developed in Davison and Goldenberg [49].

3.8 Problems

Problem 3.1 Consider a steady-state tracking problem with a desired state trajectory

$$\tilde{\xi}(t) = \begin{bmatrix} t \\ 2 \end{bmatrix}$$

and system dynamics

$$\dot{\xi} = \begin{bmatrix} 0 & 1 \\ 0 & 0 \end{bmatrix} \xi + \begin{bmatrix} 0 \\ 1 \end{bmatrix} u$$

and where $Q = \text{diag}(1, 0)$ and $R = [1]$. Reduce the tracking problem to a disturbance-rejection problem and find the optimal control input

$u^*(t) = -Kx(t) + u_{fw}(t)$ using (3.12) with T approaching ∞. Recall that $x = \xi - \tilde{\xi}$.

Problem 3.2 Consider the scalar system $\dot{x} = u + \bar{w}$, with unknown constant disturbance $\bar{w}$. For this system, design a PI controller that minimizes

$$V = \int_0^\infty (x^2 + v^2)dt$$

where $v = \dot{u}$.

Problem 3.3 Consider a steady-state disturbance-rejection problem with data

$$A = \begin{bmatrix} 0 & 1 \\ -1 & 0 \end{bmatrix};\ B = \begin{bmatrix} 0 \\ 1 \end{bmatrix};\ A_w = \begin{bmatrix} 0 & 1 \\ 0 & 0 \end{bmatrix};\ Q = \begin{bmatrix} 1 & 0 \\ 0 & 0 \end{bmatrix};\ R = [1]$$

Compute the optimal steady-state control law for this data.

Problem 3.4 Design a preview controller for a problem with performance measure

$$V = \int_t^{t+T_p} (x^2 + u^2)d\tau + Mx^2(T)$$

subject to

$$\dot{x} = 10x + u + w(t)$$

What value must M have in this case?

Problem 3.5 Use the minimax approach to find the control law that minimizes the effect of the disturbance w on the performance measure

$$V = \int_0^\infty (x^2 + u^2)dt$$

given the plant

$$\dot{x} = 10x + u + w\ ;\ x(0) = 1$$

when the disturbance satisfies

$$\int_0^\infty w^2(t)dt = 1$$

Hint: Let λ be a Lagrange multiplier and solve the minimax problem

$$\begin{aligned} \dot{x} &= 10x + u + w \\ V &= \int_0^\infty (x^2 + u^2 - \lambda w^2)dt \end{aligned}$$

Note that λ appears as a parameter in the solution for u^* and w^* (see (3.35)). The value of λ is fixed by the constraint

$$\int_0^\infty w^2(t)dt = 1$$

Problem 3.6 Consider the two-mass spring system of problem 2.13 with the disturbance signal $d(t)$ present, and with $Q = Q_1$.

1. Design a steady-state controller when $d(t)$ is a known constant equal to d_0.

2. Design a PI controller when $d(t)$ is constant but unknown. Pick $D = [1\ \ 0\ \ 0\ \ 0]$ and $R_v = [1]$.

3. Simulate your two controllers designed in (1) and (2) and illustrate their disturbance-rejection capabilities by plotting the output $Dx(t)$ of the closed-loop system with $x(0) = 0$ and $d_0 = 1$.

Problem 3.7 Consider the problem of asymptotically rejecting a disturbance signal

$$w(t) = w_0 + w_1 t$$

where w_0 and w_1 are unknown vectors (see (3.7)). Formulate and solve the state-feedback disturbance-rejection problem by generalizing the PI controller. State all necessary conditions for the existence of solutions.

Chapter 4

LQR Robustness Properties

In this chapter we explore the robustness properties of the optimal **state-feedback LQR solution**. **Kalman's inequality** is derived and applied to the robust-analysis problem. It is shown that the optimal LQR solution has some very strong robustness properties; in particular, at each channel of the plant input there is an infinite increasing gain margin and a phase margin of plus or minus 60 degrees.

4.1 Robust-Stability Condition

The **multivariable Nyquist stability criterion** (e.g., [199]) states that the stability of a closed-loop system with **loop transfer function (LTF)** matrix $H(s)$ is determined by the number of encirclements of the origin by the function $det[I + H(j\omega)]$ as ω varies from minus infinity to plus infinity (or equivalently, the number of encirclements of the -1 point by the function $det[I+H(j\omega)]-1$). The matrix $I + H(s)$ is generally referred to as the **return-difference matrix**. The term comes from the fact that $I+H = I-(-H)$, where $-H$ represents the transfer-matrix function around the loop in the presence of a subtraction block. In particular, when state-feedback is used, i.e., $u = -Kx$, and the return-difference matrix is computed at point $\times$ in Figure 4.1, $H(s)$ is given by

$$H(s) = K(sI - A)^{-1}B \tag{4.1}$$

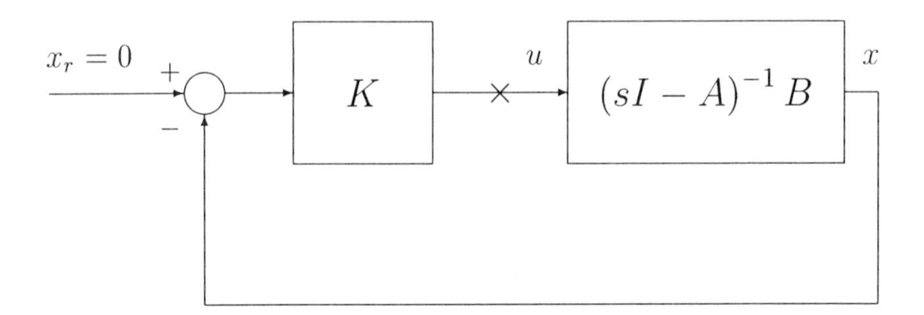

Figure 4.1: State-Feedback Configuration

Based on the multivariable Nyquist stability theorem and the **small-gain theorem**, we derive in this section a robust-stability condition for plant perturbations at the input to the plant, following the approach in [118]. It will then be shown, using Kalman's inequality, that the optimal state-feedback LQR solution has an LTF matrix $H(s)$ with very strong robustness properties. It should be noted, however, that these robustness properties are a bit serendipitous, since robustness was not part of the original LQR statement. Also, it should be noted that for these robustness properties to hold, it is critical that the state of the system be available for feedback. As will be seen in Chapter 7 if output, rather than state feedback is used, these strong robustness properties will be lost.

We model gain and phase perturbations at the input to the plant by inserting a constant matrix-transfer function L between the point $\times$ in Figure 4.1 and the plant input u. The unperturbed case (nominal system) is then represented by $L = I$. We obtain the following robust-stability result, which will be used to determine the optimal LQR stability margins.

Robust-Stability, Constant-Input Perturbation. *If the nominal closed-loop system is stable and the input-perturbation matrix L is nonsingular, then the perturbed closed-loop system will remain stable if and only if L satisfies the inequality*

$$\overline{\sigma}(L^{-1} - I) < \underline{\sigma}(I + H(j\omega)), \quad \forall \omega \tag{4.2}$$

where $\overline{\sigma}(W)$ denotes the largest singular value and $\underline{\sigma}(W)$ denotes the smallest singular value of any matrix W.

Proof: First note the identity

$$I + HL = \left[(L^{-1} - I)(I + H)^{-1} + I\right](I + H)L \tag{4.3}$$

Since the nominal feedback system is assumed stable, the term $(I + H)$ in (4.3) has a determinant with the correct number of encirclements of the origin for the Nyquist stability criterion. Also, since L is assumed constant and nonsingular it does not affect the number of encirclements. If now a condition can be placed on the square-bracketed term in (4.3) that prevents its determinant from becoming zero, then the term $det(I + HL)$ will have the correct number of encirclements for closed-loop stability. We now use the fact (small-gain theorem, see Boyd and Barratt [33], section 5.4.2) that if $\overline{\sigma}(M) < 1$ the matrix $I + M$ cannot be singular; hence, it must have a nonzero determinant. Recall that $\overline{\sigma}(M)$ is an induced norm on the matrix M; hence, $\overline{\sigma}(AB) \leq \overline{\sigma}(A)\overline{\sigma}(B)$. Thus, the small-gain theorem yields a sufficient condition for robust stability if the inequality

$$\overline{\sigma}(L^{-1} - I)\overline{\sigma}[(I + H(j\omega))^{-1}] < 1 \tag{4.4}$$

is satisfied for all ω. From the property that $\overline{\sigma}(A^{-1}) = 1/\underline{\sigma}(A)$ it then follows that condition (4.2) is a sufficient condition for robust stability. Necessity follows from a construction of a destabilizing L when condition (4.2) is not satisfied. We omit the proof of necessity here, but such a proof may be found in [199]. ■

In the next section, we show that when K is the optimal steady-state LQR feedback gain we have the (Kalman) inequality

$$1 \leq \underline{\sigma}(I + H(j\omega)), \text{ for all } \omega$$

which then implies that the inequality

$$\overline{\sigma}(L^{-1} - I) < 1 \tag{4.5}$$

will guarantee the satisfaction of condition (4.2). From (4.5) we then obtain stability margins for the optimal LQR problem by selecting appropriate values of the perturbation matrix L.

4.2 Kalman's Inequality

Kalman's inequality is a direct consequence of a matrix identity that is valid for any feedback matrix K that is a solution of the optimal steady-state LQR problem, i.e., $K = R^{-1}B'\bar{P}$, where $\bar{P}$ satisfies the ARE (2.45). We refer to this identity as Kalman's identity.

Kalman's Identity. *If $H(jw) = K(jwI - A)^{-1}B$ and K is the LQR optimal feedback matrix, then, for all w*

$$\begin{aligned} [I + H(j\omega)]^* \quad R \quad [I + H(j\omega)] &= R \\ + \quad [(j\omega I - A)^{-1}B]^* Q [(j\omega I - A)^{-1}B] & \end{aligned} \tag{4.6}$$

where Q, R, A, and B make up the usual data for the LQR problem and W^* denotes the complex conjugate transpose of any matrix W.

Proof: By assumption, $\bar{P}$ satisfies the ARE

$$A'\bar{P} + \bar{P}A - \bar{P}BR^{-1}B'\bar{P} + Q = 0 \tag{4.7}$$

If we add and subtract the term $s\bar{P}$ from the left side of (4.7) we obtain

$$-(-sI - A')\bar{P} - \bar{P}(sI - A) - \bar{P}BR^{-1}B'\bar{P} + Q = 0 \tag{4.8}$$

Now let $\Phi(s) = (sI - A)^{-1}$ and multiply (4.8) on the left by $B'\Phi'(-s)$ and on the right by $\Phi(s)B$. This yields, after moving the Q term to the right and changing signs

$$\begin{aligned} B'\Phi'(-s)Q\Phi(s)B &= B'\bar{P}\Phi(s)B + B'\Phi'(-s)\bar{P}B + \\ & \quad B'\Phi'(-s)\bar{P}BR^{-1}B'\bar{P}\Phi(s)B \end{aligned} \tag{4.9}$$

Next add R to both sides of (4.9) and use $K = R^{-1}B'\bar{P}$ and the substitution $H(s) = K(sI - A)^{-1}B = K\Phi(s)B$ to obtain

$$\begin{aligned} R &+ RH(s) + H'(-s)R + H'(-s)RH(s) = \\ R &+ \left[(-sI - A)^{-1}B\right]' Q \left[(sI - A)^{-1}B\right] \end{aligned} \tag{4.10}$$

which is precisely the multiplied-out version of (4.6) when s is replaced by $j\omega$. ■

Since the second term on the right-hand side of (4.6) is (Hermitian) positive-semidefinite, Kalman's identity implies the following matrix inequality

$$[I + H(j\omega)]^* R[I + H(j\omega)] \geq R$$

which when specialized for $R = \rho I, \rho > 0$, becomes

$$[I + H(j\omega)]^* [I + H(j\omega)] \geq I$$

Finally, using the fact that $W^*W \geq I$ is equivalent to $\underline{\sigma}(W) \geq 1$ for any matrix W, we obtain **Kalman's inequality**

$$\underline{\sigma}(I + H(j\omega)) \geq 1, \text{ for all } \omega \tag{4.11}$$

which is valid for any LTF $H(s)$ that corresponds to an **optimal LQR state-feedback control law**.

Kalman [101] first derived inequality (4.11) for the single-input case in 1964, in connection with the rather unrelated problem of determining when a given state-feedback control law is optimal with respect to some LQR problem. This is referred to as the **inverse optimal problem**. For the single-input case, (4.11) becomes

$$|(1 + H(j\omega))| \geq 1, \text{ for all } \omega \tag{4.12}$$

In [101] it is shown that (4.12) is also a sufficient condition for a given control law $u = -Kx$ to be optimal with respect to some LQR problem, i.e., if (4.12) is satisfied then matrices Q and R exist for which the feedback matrix K is optimal. In the single-input case Kalman's inequality provides a great deal of direct insight into the robustness properties of the optimal state-feedback solution. E.g., it follows from the inequality that *the Nyquist plot of the loop gain $H(j\omega)$ must lie outside a unit circle centered at the minus 1 point.* This immediately implies that the optimal system has an increasing gain margin of infinity, a decreasing gain margin of 1/2, and a phase margin of plus or minus 60 degrees. In addition, if we define the sensitivity function in the usual way, i.e., $S = (1 + H)^{-1}$, the inequality states that one has sensitivity improvement at **all frequencies**, i.e.,

$$|S(j\omega)| \leq 1, \text{ for all } \omega$$

This sensitivity result may seem surprising, since sensitivity improvement in one frequency range generally means sensitivity degradation in some other range. Indeed, this is mathematically the case if the loop-transfer function $H(s)$ has a relative degree (degree of denominator polynomial minus degree of numerator polynomial) equal to or greater than two. In particular, in this case one has Bode's integral constraint

$$\int_0^\infty ln|S(j\omega)|d\omega = 0$$

which clearly excludes the inequality on the magnitude-of-sensitivity function $S(j\omega)$ (see [113], p. 440, for further discussion of the above integral constraint). Thus, Kalman's inequality demonstrates, in an indirect way, that *the optimal loop gain* $H(s) = K(sI - A)^{-1}B$ *must have a relative degree exactly equal to one* (of course, the term $K(sI - A)^{-1}B$ has a relative degree of at least one). The cause of this "unusual" situation is the fact that the optimal controller is a static function of the system state. Such controllers result in poor performance when high-frequency sensor noise is present. As will be seen in Chapter 6, some of these problems can be overcome with dynamic output feedback.

In stating Kalman's inequality we specialized to the case where $R = \rho I$. This assumption is not necessary, but it does provide the simplest statements of robustness properties, so it will be taken as a standing assumption for the rest of our discussion on LQR robustness. We now pursue the implications of LQR optimality to robustness in the multi-input case. This is done in the next section. First, however, we explore some other important implications of Kalman's identity.

Other Implications of Kalman's Identity.

- **Cheap control**. If we factor Q in the usual way, i.e., $Q = D'D$, and we replace R by ρI, the identity (4.6) may be written

$$[I + H(j\omega)]^*[I + H(j\omega)] = \\ I + 1/\rho \left[D(j\omega I - A)^{-1}B\right]^* \left[D(j\omega I - A)^{-1}B\right] \quad (4.13)$$

  with $H(s) = K(sI - A)^{-1}B$. For cheap control ρ approaches zero, and the only way the above identity can be satisfied in the limit is that the feedback-gain matrix behave asymptotically as

$$K = \tilde{K}/\sqrt{\rho}$$

where $\tilde{K}$ is a matrix that is independent of ρ. Now the LTF matrix $H(s)$ cannot have any finite zeros in the right-half s-plane since, with all the usual assumptions, the closed-loop system is known to be stable, and with ρ approaching zero the closed-loop poles (zeros of $det(I+H(s))$) approach the zeros of $H(s)$, by root locus arguments. Actually, the fact that *for an LQR optimal K the LTF matrix $K(sI-A)^{-1}B$ cannot have any unstable zeros, i.e., must be minimum phase,* is of important independent interest. In any case, the only way that (4.13) can be satisfied for ρ approaching zero is for $\tilde{K}(sI-A)^{-1}B$ to approach $UD(sI-A)^{-1}B$, where U is an arbitrary orthogonal matrix, i.e., $U'U = I$. Since U is a nonsingular constant matrix this means that $K(sI-A)^{-1}B$ and $D(sI-A)^{-1}B$ must have the same zeros asymptotically, i.e., $D(sI-A)^{-1}B$ must be minimum phase. This is the condition placed on the cheap-control problem in Section 2.7 for the performance measure to go to zero. Note in particular that for the ARE, (4.7), to hold as ρ approaches zero, $\bar{P}$ must behave as

$$\bar{P} = \sqrt{\rho}\tilde{P}$$

where $\tilde{P}$ is independent of ρ.

- **Solution for K by spectral factorization.** Note that everything on the right side of Kalman's identity is known, i.e., Q, R, A, and B. The only "unknown" is K, which appears on the left side of the identity. In the single-input case this can be easily exploited to solve for the feedback gain K directly, without solving Riccati equations, by spectral factorization of a polynomial. Let $\Phi(s) = (sI-A)^{-1}$ be written as

$$\Phi(s) = M(s)/\phi(s)$$

where $\phi(s)$ is the characteristic polynomial of A, $\phi(s) = det(sI-A)$, and $M(s)$ is the adjunct matrix of $(sI-A)$, which can always be written

$$M(s) = s^{n-1}M_1 + s^{n-2}M_2 + \ldots + M_n$$

where $M_1 = I$. The matrices M_i and the polynomial $\phi(s)$ may be computed from Faddeeva's algorithm (see, e.g., [113], p. 34).

Kalman's identity may then be written, for general s, as the scalar equation

$$\begin{aligned}[\phi(-s) + k'M(-s)b]'[\phi(s) + k'M(s)b] &= b'M'(-s)QM(s)b \\ &+ \phi(-s)\phi(s) \quad (4.14)\end{aligned}$$

where for the single-input case K is a $1 \times n$ matrix, written $K = k'$; B is an $n \times 1$ matrix, written $B = b$; and R is taken as the 1×1 matrix, $R = [1]$. No loss of generality results from this special choice of R, since any weight on R can always be divided out and included in Q. Now let the right side of (4.14) be factored (spectral factorization) as a product of two polynomials $\psi(-s)\psi(s)$, where $\psi(s)$ has all its zeros with negative real parts. This is always possible since the right side of (4.14) is an even polynomial that is non-negative on the $j\omega$ axis. With this factorization (4.14) implies

$$\phi(s) + k'M(s)b = \psi(s) \quad (4.15)$$

If the coefficients of s are equated in (4.15) one obtains a system of linear equations for the entries of the row vector k'.

- **Low state-weighting.** From Kalman's identity it is also possible to show that when the state-weighting matrix is of the form ρQ and ρ approaches zero, the closed-loop eigenvalues approach the stable open-loop eigenvalues, and the negative (polar image) of the unstable open-loop eigenvalues. We will not demonstrate this result here. A detailed discussion may be found in chap. 6 of [4].

We illustrate the inverse-control problem and solution via the following example.

Example 4.1 Consider a single-input system with A and b matrices given by

$$A = \begin{bmatrix} 0 & 1 \\ 0 & 0 \end{bmatrix}; \; b = \begin{bmatrix} 0 \\ 1 \end{bmatrix}$$

For what values of q is the state-feedback control gain

$$k' = [1 \;\; q]$$

optimal for some LQR problem?

Solution: First, for this example stability of the control law requires that $q > 0$. Next $H(s)$ is given by

$$H(s) = \frac{1 + qs}{s^2}$$

and Kalman's identity, a necessary and sufficient condition for LQR optimality for a stable control law, requires the satisfaction of inequality (4.12) or what is equivalent

$$|1 + H(j\omega)|^2 \geq 1$$

By multiplying this inequality by ω^4 one obtains finally the requirement

$$1 + (q^2 - 2)\omega^2 \geq 0$$

which can be satisfied for all ω *if and only if* $q \geq \sqrt{2}$. Thus, for example, the stable gain matrix $k_a' = (1 \ \ 2)$ is optimal with respect to some LQR problem, but the stable gain matrix $k_b' = (1 \ \ 1)$ is not optimal with respect to *any LQR problem.*

△

Example 4.2 Consider the problem of computing the optimal feedback matrix for an LQR problem with the following data

$$A = \begin{bmatrix} 1 & 1 \\ 0 & 1 \end{bmatrix}; B = \begin{bmatrix} 0 \\ 1 \end{bmatrix}; Q = \rho \begin{bmatrix} 1 & 1 \\ 1 & 1 \end{bmatrix}; R = [1]$$

Solution: This is a single-input problem, taken from [56], that can be solved by using Kalman's identity and spectral factorization. For this example we have

$$\begin{aligned} \Phi(s) &= (sI - A)^{-1} \\ &= \frac{1}{(s-1)^2} \begin{bmatrix} s-1 & 1 \\ 0 & s-1 \end{bmatrix} \end{aligned}$$

so that we have $\phi(s) = (s-1)^2$ and

$$M(s) = \begin{bmatrix} s-1 & 1 \\ 0 & s-1 \end{bmatrix}$$

From this data the right-hand side of (4.14) becomes

$$s^4 - (2+\rho)s^2 + 1 = \psi(-s)\psi(s)$$

where $\psi(s)$ is a stable polynomial that can be written in this case as $\psi(s) = s^2 + as + 1$, where a can be computed by equating coefficients of powers of s in the above equation, with the result that $a = \sqrt{4+\rho}$. We now use identity (4.15) to compute the row vector $k' = (k_1 \;\; k_2)$. For this example we have

$$\phi(s) + k'M(s)b = s^2 + (-2+k_2)s + (1+k_1-k_2)$$

If this is equated to $\psi(s)$ and coefficients of s are equated, we obtain the linear equations

$$-2 + k_2 = \sqrt{4+\rho}, \;\; 1 + k_1 - k_2 = 1$$

Finally, the solution of these equations yields the optimal feedback matrix

$$k' = \alpha(1 \;\; 1), \text{ where } \alpha = 2 + \sqrt{4+\rho}$$

4.3 Gain and Phase Margins

We now use the robustness condition (4.2) and Kalman's inequality (4.11) to obtain gain and phase margins for each input channel to an optimal LQR feedback system. To this end let the perturbation matrix L be diagonal with entries l_i, then $(L^{-1} - I)$ is also diagonal and has singular values given by

$$\sigma_i(L^{-1} - I) = |1/l_i - 1| \tag{4.16}$$

Then the robust-stability condition on L, (4.5), becomes

$$|1/l_i - 1| < 1 \tag{4.17}$$

To study gain margin let l_i be real, then from (4.17) we have

$$-1 < 1/l_i - 1 < 1$$

which implies $1/2 < l_i < \infty$, which means an **infinite increasing gain margin**, and a **decreasing gain margin of 1/2 in each**

channel. To study the phase margin let $l_i = e^{j\theta_i}$, then (4.17) is equivalent to

$$\left|e^{-j\theta_i} - 1\right|^2 = (\cos\theta_i - 1)^2 + (\sin\theta_i)^2 < 1$$

which implies $\cos\theta_i > 1/2$ or finally $-60^0 < \theta_i < 60^0$, i.e., **a phase margin in each channel of** plus or minus 60 **degrees**.

4.4 MATLAB Software

We consider in this section the aircraft model of example 2.6.

Example 4.3 If the optimal feedback gain given in (2.66) is used, each of the three input channels can tolerate gain variations in the range $(0.5, \infty)$ and phase variations in the range $(-60^o, 60^o)$. On the other hand, assume that, perhaps due to sensor failures, the states x_4 and x_5 (spoiler angle and forward acceleration) are no longer available for feedback. This is equivalent to reducing the fourth and fifth columns in the K matrix given by (2.66) to zeros. Denote the new feedback matrix by K_1, then

$$K_1 = \begin{bmatrix} -0.2642 & -0.0329 & 0.9009 & 0 & 0 \\ 0.3003 & 1.0474 & 0.7179 & 0 & 0 \\ -0.9165 & -0.2750 & -2.5510 & 0 & 0 \end{bmatrix} \tag{4.18}$$

Since the eigenvalues of $A - BK_1$ are in the left-half plane (LHP) as given by

$$\begin{aligned} \lambda_1 &= -0.5732 + 2.9976i \\ \lambda_2 &= -0.5732 - 2.9976i \\ \lambda_3 &= -0.1949 + 0.6695i \\ \lambda_4 &= -0.1949 - 0.6695i \\ \lambda_5 &= -1.1104 \end{aligned}$$

K_1 is still a stabilizing compensator for the aircraft model. To study the resulting gain and phase margins, the MATLAB function **sigma** may be used to obtain plots of the singular values of $I + H(j\omega)$ versus ω. In particular, we used $sigma(Sys, w, type)$ with $C = K_1$, $type = 3$, $D = 0$ to return a plot of the singular values of $I + H(jw)$ with $H(jw) = C(jwI - A)^{-1}B + D$ over the frequency range specified

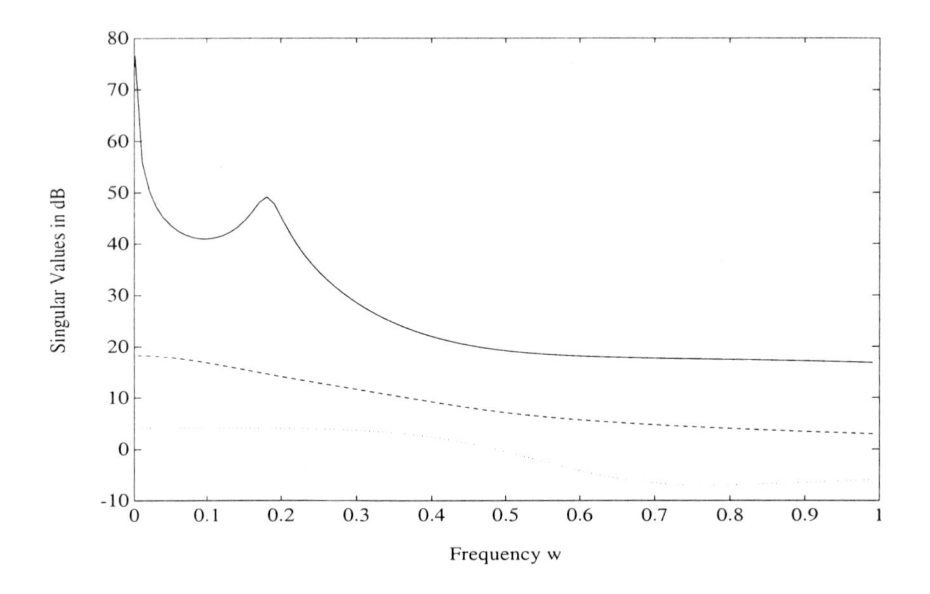

Figure 4.2: Singular Values of $I + H(jw)$ for the Aircraft Model with Sensor failures

by the vector w. The plots are shown in Figure 4.2, which from the results of problem 4.2, imply an $\alpha = 0.4457$ so that the admissible input-channel gain is in the range $(0.691, 1.804)$ and a phase margin range of $(-25.73, 25.73)$ degrees for the system with sensor failures.

4.5 Notes and References

Further discussion of gain and phase margins for the LQR problem may be found in [59], [118], and [182]. It should be noted that the term **return-ratio matrix** is sometimes used for the term **loop-transfer matrix**, e.g., in [182]. This last reference also contains detailed discussions of the role of the return-ratio and return-difference matrices in multivariable feedback systems. We have not explored other kinds of perturbations here, i.e., those beyond gain and phase margins. More general perturbations are considered in Chapter 7, where the direct design of robust systems is considered in more detail. Finally, it should be noted that optimal LQR systems also have a degree of robustness to nonlinearities (memoryless) inserted at the input point to the plant. See section 5.5 of [4], or [179] for more details on robustness with respect to nonlinearities.

4.6 Problems

Problem 4.1 Consider a single-input system with A and b matrices given by

$$A = \begin{bmatrix} 1 & 1 \\ 0 & 1 \end{bmatrix}; \quad b = \begin{bmatrix} 0 \\ 1 \end{bmatrix}$$

What are the conditions on k_1 and k_2 such that the state-feedback matrix $k' = [k_1 \;\; k_2]$ is optimal with respect to some LQR problem?

Problem 4.2 If a loop-transfer matrix $H(s)$ does not correspond to an LQR optimal system one may not be able to guarantee the stability margins given in sect. 4.3. Let

$$\inf_{\omega} \underline{\sigma}(I + H(j\omega)) = \alpha$$

where $\alpha < 1$. Use the robust condition (4.2) and the approach used in Section 4.3 to show that the gain margins and the phase margins are then

$$\begin{aligned} \frac{1}{1+\alpha} &< \; l_i < \frac{1}{1-\alpha} \\ -cos^{-1}(1-\alpha^2/2) &< \; \theta_i < cos^{-1}(1-\alpha^2/2) \end{aligned}$$

Problem 4.3 Use the spectral-factorization approach outlined in Section 4.2 to find the feedback gain for the system in problem 2.2.

Problem 4.4 Verify Kalman's inequality for the LTF matrix $H(s)$ obtained in problem 2.12, by plotting the singular values of $I + H(jw)$.

Problem 4.5 Verify that the LTF matrix $H(s)$ obtained in problem 2.2 is minimum-phase, as discussed in Section 4.2.

Problem 4.6 Repeat problem 4.5 for the aircraft example of Section 2.8.

Problem 4.7 Show that for the single-input case, Kalman's identity can be used to study the migration of the closed-loop eigenvalues as ρ is varied. In particular, show that the root locus for optimal feedback may be determined from the left-half s plane root-locus segments corresponding to the transfer function $b'\Phi'(-s)Q\Phi(s)b$. Use this result to plot the migration of the closed-loop eigenvalues for problem 2.11 part 1, when $R = [\rho]$.

Chapter 5

Stochastic Control

In this chapter we consider LQR problems in the presence of disturbance signals modeled as stochastic processes. In particular, disturbance signals are assumed to be zero-mean, Gaussian, white-noise processes. *We assume, however, that the state of the system is available for feedback, uncorrupted by any disturbance signal.* The problem of noise-corrupted measurements is considered in the next two chapters. Since the state of a system disturbed by random signals is also random, the performance measure is taken as the expected value of an integral-quadratic form. Optimization of a performance measure of this type means that we are optimizing our system performance only "on the average." Mathematically we model our dynamical system as a stochastic differential equation. With the assumptions of white-noise disturbances and state feedback, the state of the system becomes a Markov process. The principle of optimality is used to derive a "stochastic" Hamilton-Jacobi optimization equation for the Markov system. Two special cases are considered in some detail. One is the case of additive disturbances, and the other is the case of multiplicative disturbances. In the case of additive disturbances it is shown that the optimal stochastic solution is identical to the deterministic solution studied in Chapter 2. In the multiplicative case it is shown that a modified Riccati equation is required for the solution of the optimal stochastic problem.

5.1 Stochastic Differential Equations

As in the deterministic case we first generalize our stochastic linear-quadratic optimal control problem to a nonlinear time-varying problem. *In the stochastic case there is a price to be paid for this generalization; i.e., one must deal with the mathematical complexity of* **stochastic differential equations**. We treat stochastic differential equations only in a formal way and hence avoid some of the mathematical complexities. A rigorous treatment may be found in [111] and in [212]. With a formal treatment of stochastic differential equations, the derivation of the optimization equation for nonlinear dynamical systems is not significantly more difficult than for linear systems. Before proceeding to the derivation of the optimization equations, however, we need to recall some definitions and results from probability theory. In particular:

- The **correlation matrix** $R(t, \tau)$ of a zero-mean, random vector-process $z(t)$ is defined as

$$R(t, \tau) = E\{z(t)z'(\tau)\}$$

  and the **covariance matrix** $V(t)$ as

$$V(t) = E\{z(t)z'(t)\}$$

- A vector **Wiener process**, also known as **Brownian motion** or an **independent-increment process,** is a random vector-process $w(t)$ with the following properties

  1. The increments $\Delta w(t) = w(t+\Delta t) - w(t)$ are **zero-mean Gaussian**.
  2. The increments over nonoverlapping intervals are **independent**.
  3. The covariance matrix of the increments, as Δt approaches zero, is **linear** in the time interval Δt, i.e.,

$$E\{[\Delta w(t)][\Delta w(t)]'\} = W(t)\Delta t$$

  which we formally write

$$E\{dw(t)dw'(t)\} = W(t)dt$$

- Formally, the derivative of a vector Wiener process is **white noise**, also referred to as **delta-correlated noise**. The difficulty with this as a rigorous definition is that strictly speaking, time derivatives of a Wiener process do not exist. For additive disturbances in linear systems, the formal definition of the derivative of a Wiener process, $\dot{w}(t)$, as a delta-correlated process, i.e.,

$$E\{\dot{w}(t)\dot{w}'(\tau)\} = \delta(t-\tau)W$$

  where $\delta(t)$ is the impulse function, gives correct results. However, for multiplicative disturbances in linear systems or for general nonlinear systems one must go back to the properties of the increments $\Delta w(t)$.

- A **vector Markov process**, $x(t)$, is a random vector-process with the property that the probability distribution for $x(t_1)$, $t_1 \geq t$ conditioned on $x(\tau)$, $\tau \leq t$ is equal to the probability distribution conditioned only $x(t)$, the most recent value of x. Thus, for example, for a Markov process we have

$$E\{x(t_1)x'(t_1) \mid x(\tau),\ \tau \leq t\} = E\{x(t_1)x'(t_1) \mid x(t)\}$$

- **Conditional expectations** satisfy the following useful properties:

  1. **Nested expectation** $E\{x\} = E\{E\{x \mid y\}\}$ for random x and y.
  2. **Conditional certainty** $E\{f(x) \mid x\} = f(x)$.

- Two random processes, $x(t)$ and $y(t)$, are said to be **uncorrelated** if

$$E\{x(t)y'(t)\} = O$$

  where O is a matrix with zero entries. Note that if $x(t)$ and $y(t)$ are **independent zero-mean processes**, they are uncorrelated.

Details on the above definitions and results may be found in books on stochastic processes or in most books on stochastic control (e.g., [111]).

We can now proceed with a discussion of our stochastic dynamical model. We wish to consider nonlinear time-varying systems of the form

$$\dot{x} = f(x,u,t) + G(x,u,t)\dot{w}(t) \tag{5.1}$$

where $w(t)$ is a Wiener process, i.e., $\dot{w}(t)$ is a white-noise process. However, one cannot define the stochastic properties of nonlinear systems of the form given in (5.1) driven with white noise (while it is possible to do so for "additive linear systems", i.e., systems where $f(x, u, t)$ is linear in x and u, and $G(x, u, t)$ is a matrix independent of x and u). The only way to interpret equations of this type is in terms of the evolution of the differential of the state x, i.e., dx, and the differential of the Wiener process w, i.e., dw. Thus, instead of (5.1) we consider the stochastic equation of differentials

$$dx = f(x, u, t)dt + G(x, u, t)dw \tag{5.2}$$

which is commonly referred to as a **stochastic differential equation**. If the matrix $G(x, u, t)$ is independent of x and u, we say that we have an **additive-noise problem**. If $G(x, u, t)$ depends on x or u, or both, we say we have a **multiplicative-noise problem**. In particular, a multiplicative problem in x alone is often referred to as a **state-dependent noise problem** (e.g., [211]). State-dependent noise problems may be used to model uncertainty in systems parameters. We limit our discussion in this chapter to additive and state-dependent problems, but since in the derivation of the optimization equation, no special difficulty is incurred in considering the more general multiplicative-noise case, we will continue to write $G(x, u, t)$.

From the theory of stochastic differential equations (e.g., [134] or [111]), it is known that when the control input is a function of the current state and time, i.e., $u(t) = \phi(x, t)$, *then the state $x(t)$ given as a solution of (5.2) is a Markov process.* This is the key result required for the validity of a principle of optimality for stochastic systems.

5.2 Stochastic Hamilton-Jacobi Equation

We now define a stochastic optimal control problem and derive a necessary condition for optimality (stochastic Hamilton-Jacobi equation). Since for random disturbances the state is random, we must use some statistical property of the state to formulate a performance measure that is a real, nonrandom variable. The statistic most commonly selected is the **expected value**; thus, we define the following performance measure for the stochastic control problem

$$V = E\{\int_t^T l(x(\tau), u(\tau), \tau)\, d\tau + m(x(T)) \mid x(t) = x\} \tag{5.3}$$

We assume, as usual, that T is fixed and that $l(x, u, t)$ and $m(x)$ have the same properties as in the deterministic case. Note that with the conditioning on $x(t)$, V becomes a function of the initial state x in addition to the initial time t, i.e., $V = V(x, t)$. The optimal stochastic control problem considered here is then to minimize V given in (5.3) with respect to the control input u, subject to the constraint that the state $x(\tau)$ satisfy the stochastic differential equation

$$dx = f(x, u, t)dt + G(x, u, t)dw \tag{5.4}$$

where $dw(t)$ is a Wiener increment, with covariance matrix $W(t)dt$.

We now state a stochastic optimality principle that will be used to develop a necessary condition and an optimal state-feedback solution.

Stochastic Optimality Principle. *If $u^*(\tau)$ is optimal over the interval $[t, T]$, conditioned on the initial state $x(t)$, then $u^*(\tau)$ is necessarily optimal over the subinterval $[t + \Delta t, T]$ for any Δt such that $T - t \geq \Delta t > 0$.*

Proof: With the Markov property of $x(\tau)$ and the conditioning in the performance measure, it follows that the performance-measure value over the subinterval $[t+\Delta t, T]$ conditioned on $x(t+\Delta t)$ is completely independent of the value of $u(\tau)$ over the interval $[t, t + \Delta t]$. This is really all that is needed to make the deterministic proof of the optimality principle apply to the stochastic problem. We will not repeat all the inequalities required to complete the proof but will simply note that with appropriate conditional expectations, all the inequalities in the deterministic proof given in Section 2.2 hold.

■

We now derive an optimization equation for the above stochastic control problem. First, we denote by $V^*(x, t)$ the optimal value of performance measure V given by (5.3). Note that V^* depends on the initial state $x(t) = x$ because of the conditioning on the expected value. As in the deterministic case, we denote by $u[t, T]$ the control input defined over the interval $[t, T]$. Also for convenience we suppress the arguments of the functions $l(x, u, t), m(x), f(x, u, t)$, and $G(x, u, t)$ in the expressions that follow. We have from the definition of V^*

$$V^*(x,t) = \min_{u[t,T]} E\left\{\int_t^{t+\Delta t} l d\tau + \left(\int_{t+\Delta t}^{T} l d\tau + m\right) \mid x(t)\right\} \quad (5.5)$$

If we use the **nested-expectation property**, we can then write

$$V^*(x,t) = \min_{u[t,T]} E\{\int_t^{t+\Delta t} l d\tau + E\{\left(\int_{t+\Delta t}^{T} l d\tau + m\right) \mid x(t+\Delta t)\} \mid x(t)\} \quad (5.6)$$

Note that in (5.5) and (5.6) we have written the conditioning directly in terms of the state in question, e.g., $x(t)$, rather than using the more precise notation $x(t) = x$ simply for convenience. If we now invoke the stochastic optimality principle, (5.6) becomes

$$V^*(x,t) = \min_{u[t,t+\Delta t]} E\left\{\int_t^{t+\Delta t} l d\tau + V^*(x(t+\Delta t), t+\Delta t) \mid x(t)\right\} \quad (5.7)$$

Note now that $x(t+\Delta t)$, which appears as an argument of V^* in (5.7), is a random vector given by $x(t+\Delta t) = x + \Delta x$, where, from the stochastic differential equation (5.4), Δx is given approximately by

$$\Delta x = f\Delta t + G\Delta w \quad (5.8)$$

We use next a multivariable Taylor-series expansion of V^* about the point (x,t) and an approximation of the integral in (5.7) to obtain our final optimization equation. *Because of the fact that the covariance of a Wiener process is* **linear** *in Δt we must use the Taylor-series expansion up to quadratic terms in Δx.* In particular, we need the following terms in the Taylor series for V^*

$$\begin{aligned} V^*(x+\Delta x, t+\Delta t) &= V^*(x,t) + \frac{\partial V^*}{\partial t}\Delta t \\ &+ \left[\frac{\partial V^*}{\partial x}\right]' \Delta x + \frac{1}{2}(\Delta x)' V_{xx} (\Delta x) \end{aligned} \quad (5.9)$$

where the matrix V_{xx} is used to denote a matrix whose $(i,j)^{\text{th}}$ entry is given by

$$\frac{\partial^2 V}{\partial x_i \partial x_j}$$

With the above approximations, (5.7) becomes

$$\begin{aligned} V^*(x,t) &= \min_{u(t)} E\{l(x,u,t)\Delta t + V^*(x,t) + \frac{\partial V^*}{\partial t}\Delta t + \left[\frac{\partial V^*}{\partial x}\right]' \Delta x \\ &+ \frac{1}{2}(\Delta x)' V_{xx}(\Delta x) \mid x\} \end{aligned} \tag{5.10}$$

If we use the **conditional-certainty property** of expectations and the fact that the Wiener process increment has zero mean, we have

$$\begin{aligned} E\{l(x,u,t)\Delta t \mid x\} &= l(x,u,t)\Delta t \\ E\{\left[\frac{\partial V^*}{\partial x}\right]' (f\Delta t + G\Delta w) \mid x\} &= \left[\frac{\partial V^*}{\partial x}\right]' f(x,u,t)\Delta t \end{aligned}$$

Also, using the fact that for any vector z and symmetric matrix S we have

$$z'Sz = tr\ Szz'$$

where $tr\ N$ denotes the trace of the matrix N, and thus

$$E\{(\Delta x)' V_{xx}(\Delta x) \mid x\} = tr\ V_{xx} E\{(\Delta x)(\Delta x)' \mid x\}$$

If we now use the fact that Δw has zero mean and a covariance matrix given by $W\Delta t$, we have

$$\begin{aligned} E\{(\Delta x)(\Delta x)' \mid x\} &= E\{(f\Delta t + G\Delta w)(f\Delta t + G\Delta w)' \mid x\} \\ &= ff'(\Delta t)^2 + GWG'\Delta t \end{aligned} \tag{5.11}$$

Note that in (5.11) the computed expectation term $ff'(\Delta t)^2$ is of order two in Δt and may be neglected in the limit, compared with first-order terms such as the term $GWG'\Delta t$. Indeed, if we now take the limit as Δ approaches zero and use all of the above computations for conditioned expectations, we obtain the **stochastic Hamilton–Jacobi equation**

$$\begin{aligned} -\frac{\partial V^*}{\partial t} &= \min_u \{l(x,u,t) + \left[\frac{\partial V^*}{\partial x}\right]' f(x,u,t) \\ &+ \frac{1}{2}\, tr\ V_{xx} G(x,u,t) W G'(x,u,t)\} \end{aligned} \tag{5.12}$$

with boundary condition

$$V^*(x,T) = m(x), \quad \text{for all } x \tag{5.13}$$

The boundary condition (5.13) results directly from (5.3) and the properties of conditioned expectations. Note that if no disturbance is present, i.e., $W = 0$, the stochastic Hamilton–Jacobi equation, (5.12), reduces to the deterministic Hamilton–Jacobi equation of Chapter 2, i.e., (2.13). As in the deterministic case, (5.12) is solved in three steps. First, one computes the feedback function $u^* = \psi(\partial V^*/\partial x, x, t)$, which minimizes the bracketed term in (5.12), then one must substitute this ψ back into (5.12); and finally, one must solve the resulting partial-differential equation for $V^*(x,t)$. *Note that we have assumed in this last statement that G is independent of u, otherwise the minimization step will involve a function ψ, which also depends on V_{xx}. This will be a standing assumption in the rest of this chapter.* The nonlinear optimization equation that must be solved in the stochastic case is now a **second-order** partial differential, and, as can be imagined, solutions to general problems are even more difficult to obtain than solutions to deterministic ones. However, for an integral-quadratic performance measure, two special problems do have explicit solutions, at least in terms of solutions of Riccati equations. We consider these special cases in the next two sections.

5.3 Additive Disturbances

In this section we study a stochastic optimal control problem where the performance measure is quadratic, the system is linear, and the disturbance appears additively, i.e.,

$$V = E\{\int_t^T (x'Qx + u'Ru)d\tau + x'(T)Mx(T) \mid x\} \tag{5.14}$$

and

$$dx = (Ax + Bu)dt + Gdw \tag{5.15}$$

In terms of the functions $l(x,u,t)$, $m(x)$, $f(x,u,t)$, and $G(x,t)$, this means $l = x'Qx+u'Ru, m = x'Mx, f = Ax+Bu$, and $G(x,t) = G(t)$. In (5.15), w is a Wiener process with a differential covariance matrix equal to Wdt. The data for this optimal stochastic LQ problem are thus given by the matrices Q, R, M, A, B, G, and W. In this case, the minimization step required in the stochastic Hamilton-Jacobi equation, (5.12), involves only the minimization of the terms

$$u'Ru + \left[\frac{\partial V^*}{\partial x}\right]' Bu \tag{5.16}$$

since these are the only terms that involve u. These are precisely the same terms involved in the deterministic LQ problem in Chapter 2, where the optimal value of u was found to be

$$u^* = -\frac{1}{2}R^{-1}B'\frac{\partial V^*}{\partial x} \tag{5.17}$$

One might be tempted at this point to assume a solution for V^* of the form $V^* = x'P(t)x$ as in the deterministic case. However, this introduces a term independent of x on the right-hand side of (5.12) that would not be balanced on the left. An appropriate form is

$$V^* = x'P(t)x + c(t) \tag{5.18}$$

If u^* given by (5.17) and V^* given by (5.18) are substituted back into the stochastic Hamilton-Jacobi equation, (5.12), and use is made of the fact that $\partial x'Px/\partial x = 2Px$ and the fact that $V_{xx} = 2P$, one obtains after some matrix manipulations the equation

$$-x'\dot{P}x - \dot{c} = [x'(A'P + PA + Q - PBR^{-1}B'P)x + tr\ PGWG'] \tag{5.19}$$

If coefficients of like powers of x are equated in (5.19) we finally obtain the optimization equations

$$-\dot{P} = A'P + PA + Q - PBR^{-1}B'P \tag{5.20}$$

and

$$-\dot{c} = tr\ PGWG' \tag{5.21}$$

The boundary conditions for (5.20) and (5.21) computed from (5.13) are

$$P(T) = M, \quad \text{and} \quad c(T) = 0 \tag{5.22}$$

with the optimal state-feedback control law given by

$$u^*(t) = -R^{-1}B'P(t)x(t) \tag{5.23}$$

where $P(t)$ is given as the solution of the Riccati equation (5.20). *Note that the optimal control for this additive-disturbance stochastic* *problem is identical to the optimal control for the deterministic case, i.e., the case where $w(t) = 0$ or equivalently $W = 0$.* The deterministic case may be viewed as a special case where dw is replaced by its average (expected) value, zero. When a stochastic problem can be solved as a deterministic problem with stochastic variables replaced by their average values, we say that a **certainty-equivalence principle** holds. Certainty equivalence is commonly used in engineering design, even when its validity has not been established. The above development establishes the validity of certainty equivalence for the additive-disturbance stochastic LQR problem. It should be noted, however, that a key assumption for certainty equivalence to hold in this case is that the additive noise is an **independent-increment** process, or equivalently that $\dot{w}$ is "white." Since all of the matrices defining the problem may be time-varying, the solution given above is valid for general time-varying systems.

While the control law for additive disturbances is independent of the disturbance, the optimal performance-measure value does depend on the disturbance statistics. Indeed, from the integration of (5.21) we see that

$$c(t) = \int_t^T tr\ P(\tau)GWG\ d\tau \tag{5.24}$$

where $W dt$ is the covariance matrix for the differential dw.

We explore next the steady-state solution for the additive disturbance problem. For the steady-state problem we assume that all the matrices defining the problem are constant and that T approaches infinity. As T approaches infinity the Riccati equation (5.20) becomes a standard ARE, which with the usual assumptions of stabilizability, detectability, etc., will have a unique positive-definite solution $\bar{P}$ and will lead to a closed-loop system that is asymptotically stable. Note, however, that if $P(t)$ approaches a constant value $\bar{P}$, the performance measure approaches the value

$$V^* = x'\bar{P}x + (T - t)tr\ \bar{P}GWG' \tag{5.25}$$

and thus V^* blows up as T approaches infinity. One can keep the performance measure finite by dividing V^* given above by T, and this is generally what is done in this case. The problem is caused by the fact that the additive disturbance, dw, has a persistent effect on the state trajectories and the state never goes to zero, even if the closed-loop system is asymptotically stable. In fact, by dividing by T, the performance measure

$$V = E\{\frac{1}{T}\int_0^T l(x,u,t)dt \mid x(0) = x\} \tag{5.26}$$

becomes, in the limit as T approaches infinity,

$$V = \lim_{T\to\infty} E\{l(x,u,t)\} \tag{5.27}$$

and it is the "persistent" term $E\{l(x,u,t)\}$ that is minimized by a $T-$divided performance measure. In particular, for the stochastic LQR problem defined above, the modified optimal performance-measure value is given by

$$\begin{aligned} V^* &= \lim_{t\to\infty} E\{x'(t)Qx(t) + u'(t)Ru(t)\} \\ &= tr\ \bar{P}GWG' \end{aligned} \tag{5.28}$$

Since the Riccati equation for the additive-noise case is the same as the standard Riccati equation studied in Chapter 2, we will not say anything further about the computation of control laws for this case.

5.4 Multiplicative Disturbances

In this section we solve a stochastic LQR problem for linear dynamical systems with multiplicative disturbances, in particular, linear systems where $G(x,t)$ is a matrix of the form

$$G(x,t) = (A_1x \mid A_2x \mid \ldots \mid A_qx) \tag{5.29}$$

where $q = dim\ w$. The notation in (5.29) means that the matrix G is made up of columns given by A_ix. The linear-system equation corresponding to a G of this form is given by

$$dx = (Ax + Bu)dt + (A_1xdw_1 + A_2xdw_2 + \ldots + A_qxdw_q) \tag{5.30}$$

Where dw_i is the ith component of the Wiener increment vector dw. Note that the disturbance, or noise, enters the dynamics as a multiple of the state. For this reason, this case of multiplicative noise is often referred to as the **state-dependent noise problem** [211]. A model of the type given in (5.30) could represent a problem where the A

matrix is stochastically perturbed from its nominal value A to a perturbed value

$$A + A_1 dw_1 + A_2 dw_2 + \ldots + A_q dw_q \tag{5.31}$$

Note, however, that the B matrix is assumed to be unperturbed. A physical interpretation of the perturbation represented by (5.31) is that the A matrix of the system has structured "white-noise" parameter perturbations. In this section we limit our discussion of multiplicative disturbances to state-dependent noise problems as defined above.

The stochastic optimization problem is then to minimize a quadratic performance measure of the type given in (5.14), subject to the stochastic differential equation given in (5.30). The first step in solving this problem, i.e., performing the minimization in the stochastic Hamilton–Jacobi equation, (5.12), yields the same result as in the case of additive disturbance since no new terms in u are introduced by the state-dependent-noise assumption. Thus, the optimal control law has the form

$$u^* = -\frac{1}{2} R^{-1} B' \frac{\partial V^*}{\partial x} \tag{5.32}$$

Now a solution of the form $V^* = x'P(t)x$ does work, since with this form the term $tr\ V_{xx} G(x) W G'(x)$ is quadratic in x and no constant term is required in V^*, as in the additive case. To see that the term in question is indeed quadratic we need some properties of the trace operation. The algebra is also simplified if we assume that the matrix W is the identity matrix. There is no loss in generality in doing this when W is positive definite, since W can always be factored and incorporated in G. In any case, we need the following trace properties

$$\begin{aligned} tr(ABC) &= tr(BCA) = tr(CAB); \\ tr(A+B) &= tr(A) + tr(B) \end{aligned} \tag{5.33}$$

and the following expansion for the term $G(x,t)G'(x,t)$, where $G(x,t)$ is given by (5.29)

$$G(x,t)G'(x,t) = A_1 x x' A_1' + A_2 x x' A_2' + \ldots + A_q x x' A_q' \tag{5.34}$$

With $V^* = x'Px$, we have $\partial V^*/\partial x = 2Px$ and $V_{xx} = 2P$ and the term $\frac{1}{2} tr\ V_{xx} G G'$ becomes, after some manipulations with traces

$$\frac{1}{2} tr\ V_{xx} G(x,t) G'(x,t) = x'(A_1' P A_1 + \ldots + A_q' P A_q) x \tag{5.35}$$

From the stochastic Hamilton-Jacobi equation it then follows that $P(t)$ must satisfy the equation

$$-\dot{P} = A'P + PA + Q - PBR^{-1}B'P + \sum_{i=1}^{q} A_i'PA_i \tag{5.36}$$

with the boundary condition

$$P(T) = M \tag{5.37}$$

The state-feedback control law is given by

$$u^*(t) = -R^{-1}B'P(t)x(t) \tag{5.38}$$

where $P(t)$ is the solution of (5.36).

We refer to (5.36) as the **stochastic Riccati equation**. Note that this equation is simply the standard Riccati equation modified by the terms $A_i'PA_i$ introduced by the state-dependent noise. Let us denote this term by $\Pi(P)$, i.e.,

$$\Pi(P) = \sum_{i=1}^{q} A_i'PA_i \tag{5.39}$$

We consider next steady-state solutions of the state-dependent noise problem. Assuming the stochastic Riccati equation solution, $P(t)$, approaches a constant P, as T approaches ∞, we obtain the following algebraic equation for P, which we refer to as the **stochastic algebraic Riccati equation (SARE)**, sometimes also referred to as the **generalized algebraic Riccati equation (GARE)** [188],

$$0 = A'P + PA + Q - PBR^{-1}B'P + \Pi(P) \tag{5.40}$$

where $\Pi(P)$ is given as in (5.39). Unfortunately, the usual assumptions of stabilizability, detectability, positive definiteness, etc., are not sufficient to guarantee solutions to the SARE. The term $\Pi(P)$ in (5.40) significantly complicates the problem. It is shown in [209] that if, in addition to the usual assumptions, the condition

$$\inf_K \left\| \int_0^\infty e^{(A-BK)'t}\Pi(I)e^{(A-BK)t}dt \right\| < 1 \tag{5.41}$$

holds, where $\Pi(I)$ denotes $\Pi(P)$ evaluated at $P = I$, then by using an approximation-in-policy-space algorithm, a converging sequence for the solution of (5.40) is obtained. We will not present the details of this iterative solution here but rather will simply outline the approximation-in-policy-space algorithm below and say a few words about approximation in policy space in general in the next section.

Approximation-in-Policy-Space Solution. Let K_i be a feedback matrix that makes the matrix $A - BK_i$ stable. Stabilizability of the pair (A, B) guarantees the existence of such a "starting" K. Solve the linear matrix equation

$$0 = (A - BK_i)'P_i + P_i(A - BK_i) + Q + K_i'RK_i + \Pi(P_i) \quad (5.42)$$

for P_i. Then let

$$K_{i+1} = R^{-1}B'P_i \quad (5.43)$$

and update K_i with K_{i+1}. Condition (5.41) is sufficient to guarantee that this iteration will converge to a bounded positive-definite solution for P when $Q > 0$ [209]. It is interesting to note that the minimization in (5.41) is equivalent to a cheap-control problem (see Section 2.7), where the state-weighting matrix is given by $\Pi(I)$. The feedback gain matrix given by $K = R^{-1}B'P$, where P is a solution of (5.40), leads to a closed-loop matrix $A - BK$ that is asymptotically stable and to a closed-loop system that is "mean-square" stable in the sense that

$$\lim_{t \to \infty} E\{x'(t)x(t)\} = 0 \quad (5.44)$$

We consider next an example, suggested by Wonham [211], that illustrates that if the noise intensity in a state-dependent problem becomes too large, there may not exist any optimal solution.

Example 5.1 Consider the problem of minimizing the performance measure

$$V = E\{\int_0^\infty (x'x + u^2)dt \mid x(0)\}$$

given the stochastic differential equation

$$dx = \left[\begin{pmatrix} 0 & 1 \\ 0 & 0 \end{pmatrix} x + \begin{pmatrix} 0 \\ 1 \end{pmatrix} u\right] dt + \sqrt{\gamma}\begin{bmatrix} 1 & 0 \\ 0 & 1 \end{bmatrix} x dw_1$$

Solution: In this case, $A_1 = \sqrt{\gamma}I$ and condition (5.41) becomes

$$\gamma < \frac{1}{\inf_K \|S\|}$$

where S is given by

$$S = \int_0^\infty e^{(A-BK)'t} e^{(A-BK)t} dt$$

and S may be computed as the solution of the Lyapunov equation

$$0 = (A - BK)'S + S(A - BK) + I$$

If we let $K = [k_1 \quad k_2]$ and solve this Lyapunov equation for S we obtain

$$S = \begin{bmatrix} \frac{1}{2}\left(\frac{k_2}{k_1} + \frac{k_1}{k_2}\right) + \frac{1}{2k_2} & \frac{1}{2k_1} \\ \frac{1}{2k_1} & \frac{1}{2}\left(\frac{1}{k_1 k_2} + \frac{1}{k_2}\right) \end{bmatrix}$$

The norm of S approaches its inferior value one, as $k_1 = k_2 = \alpha$ and α approaches ∞. In this case, the minimization of the norm of S with respect to K yields a minimal norm value of one. Thus, condition (5.41) requires that γ satisfy the inequality $\gamma < 1$. It is shown in [211] that for this example, this condition is also a necessary condition for a positive-definite solution for P to exist.

5.5 Sufficient Conditions for Optimality

In this section we present a sufficient condition for optimality and outline the approximation-in-policy-space approach for stochastic control problems. First, it should be noted that, as in the deterministic case, the Hamilton-Jacobi equation without the minimization operation can be used to evaluate the performance measure for a given control law. Thus, e.g., given the control law $u(t) = \phi(x,t)$ the corresponding value of V, denoted V^ϕ, satisfies the equation

$$\begin{aligned} -\frac{\partial V^\phi}{\partial t} &= l(x,\phi,t) + \left[\frac{\partial V^\phi}{\partial x}\right]' f(x,\phi,t) \\ &\quad + \frac{1}{2}\, tr\; V^\phi_{xx} G(x,\phi,t) W G'(x,\phi,t) \end{aligned} \tag{5.45}$$

with boundary condition

$$V^\phi(x,T) = m(x), \quad \text{for all } x \tag{5.46}$$

The proof that V^ϕ must satisfy (5.45) and (5.46) is similar to the deterministic proof, i.e., the additive property of integrals is used together with a Taylor-series expansion, etc. However, the proof that solutions of (5.45) with boundary condition (5.46) yield the performance measure value V^{ϕ} is more complex (see [111]) and will not be presented here. In any case, (5.45) is of independent interest in analyzing given control laws. Use will be made of this equation in the next section to develop some important duality properties between deterministic and stochastic control problems.

We state next a sufficient condition for optimality for the case where G is independent of u. *If $V^{\phi}(x,t)$, corresponding to a given control law $u = \phi(x,t)$, is continuous in all its arguments and has the necessary derivatives with respect to x and t, then the satisfaction of the inequality*

$$l(x,\phi,t) + \left[\frac{\partial V^{\phi}}{\partial x}\right]' f(x,\phi,t) \leq l(x,u,t) + \left[\frac{\partial V^{\phi}}{\partial x}\right]' f(x,u,t) \quad (5.47)$$

for all x and u, is sufficient to guarantee that the control law $u^* = \phi(x,t)$ is optimal.

Proof: We give here only a brief sketch of the proof. The proof is similar to the deterministic case; however, we must now use the fact that the performance-measure value satisfies the second-order partial-differential equation (5.45). We must now add the term

$$\frac{\partial V^{\phi}}{\partial t} + \frac{1}{2} tr\{V^{\phi}_{xx} G(x,t) G'(xt)\}$$

to both sides of (5.47) and integrate from t to T. Then we use the fact that

$$V^{\phi}(x,t) = E\{\int_t^T l(x,\phi,t)d\tau + m(x(T)) \mid x(t) = x\} \quad (5.48)$$

to obtain finally

$$V^{\phi} \leq E\{\int_t^T l(x,u,t)d\tau + m(x(T)) \mid x(t) = x\} \quad (5.49)$$

Equation (5.49) states that $u^* = \phi$ yields a smaller value of V than any other control u. ■

For the case where G is independent of u the sufficient condition (5.47) appears to be identical to the sufficient condition of Section

2.6 for deterministic systems. However, the difference is that in the stochastic case V^ϕ must satisfy the second-order partial-differential equation (5.45). When a positive-definite solution P exists that satisfies (5.40) then $V^* = x'Px$ can be used to demonstrate the optimality of the control law $u^* = -R^{-1}B'Px$ for the state-dependent-noise case. Similarly, $V^* = x'Px + c$ can be used to demonstrate optimality in the additive-noise case.

Stochastic Approximation in Policy Space. The stochastic condition for approximation in policy space is similar to the deterministic condition given in Section 2.6, with the same observation on the computation of V^i for any given optimal approximation u_i. We will not repeat the general statement of the approximation-in-policy-space algorithm given in Section 2.6 here but will simply note that for the stochastic LQR problem with state-dependent noise, (5.45) requires the solution of (5.42) to compute $V^i = x'P_ix$ for $u = -K_ix$ and K_{i+1} must equal $R^{-1}B'P_i$ for the policy improvement.

5.6 Stochastic-Deterministic Dualism

It turns out that certain stochastic problems are mathematically similar to certain deterministic problems. This dualism between two different domains can be very valuable and indeed will be exploited in the derivation of optimization equations for state estimation considered in the next chapter. In particular, what we would like to show is that the equation required to compute the value of V for

$$V_d = \int_0^\infty x'Qxdt \tag{5.50}$$

given the deterministic system

$$\dot{x} = Ax, \quad x(0) \tag{5.51}$$

is "similar" to the equation required to compute the value of V_s for

$$V_s = \lim_{t\to\infty} E\{x'cc'x\} \tag{5.52}$$

given the stochastic system

$$dx = Axdt + dw \tag{5.53}$$

where dw is a Wiener process with covariance matrix $W dt$.

From (2.49) we know that the value of V_d in the **deterministic** case is given by (for A stable)

$$\begin{aligned} V_d &= tr\ Px(0)x'(0); \\ 0 &= A'P + PA + Q \end{aligned} \tag{5.54}$$

On the other hand, from (5.45) and the steady-state results of the additive-noise case we know for the stochastic problem that

$$\begin{aligned} V_s &= tr\ SW; \\ 0 &= A'S + SA + cc' \end{aligned} \tag{5.55}$$

As is shown below, V_s given by (5.55) is identical to

$$\begin{aligned} V_s &= tr\ Scc'; \\ 0 &= AS + SA' + W \end{aligned} \tag{5.56}$$

Note that this step simply interchanges cc' and W in (5.55). If we now compare (5.54) with (5.56) we see that the deterministic-analysis problem is identical to the stochastic-analysis problem with the following (dual) substitutions:

- Replace the initial state $x(0)$ by its dual c.
- Replace the state-weighting matrix Q by its dual W, the "noise-intensity matrix."
- Replace the system matrix A by its dual A'.

With these replacements the deterministic solution can be used to solve a stochastic problem. To show that $tr\ SW$, where S is given by (5.55), is the same as $tr\ Scc'$, where S is given by (5.56), consider the following solution for $tr\ SW$, valid for any stable A

$$V = tr\ SW = tr \int_0^\infty e^{A't} cc' e^{At} dt\ W$$

If one uses the fact that $tr\ FGH = tr\ GHF = tr\ HFG$, the above integral can also be written

$$V = tr \int_0^\infty e^{At} W e^{A't} dt\ cc'$$

which is precisely the solution for (5.56).

The following simple example illustrates stochastic-deterministic dualism.

Example 5.2 Consider the stochastic problem of computing

$$\lim_{t\to\infty} E\{x'(t)cc'x(t)\}$$

where $c' = (1 \;\; 0)$ for the system

$$dx = \begin{bmatrix} 0 & 1 \\ 0 & 0 \end{bmatrix} x\, dt + dw$$

where dw has a covariance matrix equal to Idt.

Solution: This problem is identical to the deterministic problem of computing

$$\int_0^\infty x'Ixdt$$

for the system

$$\dot{x} = \begin{bmatrix} 0 & 0 \\ 1 & 0 \end{bmatrix} x; \;\; x(0) = c = \begin{bmatrix} 1 \\ 0 \end{bmatrix}$$

5.7 MATLAB Software

Since the Riccati equation for the additive-noise case is the same as the deterministic equation, the standard **lqr** function can be used to solve steady-state additive-noise problems. There is no specific software to solve the stochastic algebraic Riccati equation. The best one can do with existing software is to write a system of linear equations for the linear approximation-in-policy-space equation (5.42) and to solve the state-dependent problem by iteration. We illustrate this approach in the following example.

Example 5.3 Helicopter Model. Consider the linearized model of a CH-47 tandem-rotor helicopter in horizontal motion about a nominal airspeed of 40 knots, as discussed in [59]. The model parameters are given by

$$\begin{aligned} \dot{x} &= Ax + Bu + G\dot{w} \\ y &= Cx \end{aligned}$$

where

$$A = \begin{bmatrix} -0.02 & 0.005 & 2.4 & -32 \\ -0.14 & 0.44 & -1.3 & -30 \\ 0 & 0.018 & -1.6 & 1.2 \\ 0 & 0 & 1 & 0 \end{bmatrix}$$

$$B = \begin{bmatrix} 0.14 & -0.12 \\ 0.36 & -8.6 \\ 0.35 & 0.009 \\ 0 & 0 \end{bmatrix}$$

$$C = \begin{bmatrix} 0 & 1 & 0 & 0 \\ 0 & 0 & 0 & 57.3 \end{bmatrix}$$

The incremental outputs are

- y_1 is the vertical velocity (knots/hr)
- y_2 is the pitch altitude (radians)

and the inputs are

- u_1 is the collective rotor thrust
- u_2 is the differential collective rotor thrust

We assume next that additive white noise appears in each input channel so that $G = B$. The white-noise signal $\dot{w}$ has a covariance matrix $W = \sigma I_{2\times 2}$, where the parameter σ is introduced to study the effect of various noise levels. The performance-index matrices are given by $Q = C'C$ and $R = I_{2\times 2}$. The final time is $T = \infty$. Since for the additive-noise case the computation of the feedback matrix K is similar to the deterministic case, we use the MATLAB function **lqr** to compute K. We are also interested in the steady-state, mean-squared error due to the white-noise disturbance signal and given by

$$V^* = \lim_{t\to\infty} E\{x'(t)Qx(t) + u'(t)Ru(t)\}$$

or

$$V^* = tr\{PGWG'\} = \sigma tr\{PBB'\}$$

where P is the solution to the ARE

$$0 = A'P + PA + Q - PBR^{-1}B'P$$

Therefore, both P and K are obtained from the MATLAB command $[K, P] = lqr(A, B, Q, R)$, which for this problem leads to

$$K = \begin{bmatrix} -0.0033 & 0.0472 & 14.6421 & 60.8894 \\ 0.0171 & -1.0515 & 0.2927 & 3.2469 \end{bmatrix}$$

$$P = \begin{bmatrix} 0.0071 & -0.0021 & -0.0102 & -0.0788 \\ -0.0021 & 0.1223 & 0.0099 & -0.1941 \\ -0.0102 & 0.0099 & 41.8284 & 174.2 \\ -0.0788 & -0.1941 & 174.2 & 1120.9 \end{bmatrix}$$

Moreover, $V^* = 14.18\sigma$.

5.8 Notes and References

The texts of Anderson and Moore [4], Kwakernaak and Sivan [113], Maybeck [134], and Stengel [191] contain more details on the stochastic additive-noise case. One of the early papers, from 1961, on the control of the type of Markov systems discussed in this chapter may be found in Florentine [64]. Critical issues in the "white-noise" assumption for disturbances are the physical meaning of such models and the proper mathematics to deal with this idealization. The article by Wong and Zakai [208] contains an early discussion of some of these issues. A careful development of the control of systems characterized by stochastic differential equations may be found in the text of Kushner [112] as well as Arnold [7] and Caines [37].

The steady-state state-dependent-noise case was first considered in detail in Wonham [211]. Actually, in this reference both additive- and multiplicative-noise terms were included in the system dynamics. More detailed studies of the stochastic algebraic Riccati equation may be found in de Souza and Fragoso [188] and Wonham [209]. The issue of stability of stochastic differential equations is rather complex. In this chapter we have introduced only the concept of "mean-square" asymptotic stability, but many other definitions are possible. Further details on stochastic stability may be found in Kushner [112]. Since the state-dependent case may be used as a model of structured-parameter perturbations, an interesting question that arises is, What deterministic robustness properties does one have for the optimal state-dependent solution? A discussion of this issue may be found in Bernstein [22]. Finally, the design of fixed-order compensators for state-dependent noise problems is considered in Bernstein and Hyland [26].

We limited our discussion in this chapter to problems where the random signals multiply only the state of the system and not the input. For a discussion of problems with "control-dependent noise" see Kleinman [109]. For other references that consider state-dependent or control-dependent stochastic problems see: [28], [86], [87], [93], [135], [168], and [207].

5.9 Problems

Problem 5.1 Consider the following system with additive white-noise disturbance $\dot{w}$

$$\dot{x} = x + u + \dot{w}$$

Compute the minimal value of

$$V = \lim_{t \to \infty} E\{x(t)^2 + u(t)^2\}$$

possible, given the correlation function $10\delta(t - \tau)$ for $\dot{w}$.

Problem 5.2 As in the example of Section 5.4, consider the problem of minimizing the performance measure

$$V = E\{\int_0^\infty (x'x + u^2)dt \mid x(0)\}$$

given the stochastic differential equation

$$dx = \left[\begin{pmatrix} 0 & 1 \\ 0 & 0 \end{pmatrix} x + \begin{pmatrix} 0 \\ 1 \end{pmatrix} u\right] dt + \sqrt{\gamma} \begin{bmatrix} 0 & 0 \\ 1 & 0 \end{bmatrix} x dw_1$$

Verify that condition (5.41) is satisfied for any positive value of γ. Solve (5.40) directly for P, when $\gamma = 10$. Note: You may have to find the roots of a fourth-order polynomial to compute the entries of P.

Problem 5.3 For the data in problem 5.2, use the approximation-in-policy-space algorithm to find a control law that yields improved performance over the control law $u = -k'x$, where $k' = [8 \;\; 4]$.

Problem 5.4 Consider the two-mass spring system described in problem 2.13. Let $d = \dot{w}$ be a white-noise signal with $E\{\dot{w}(t)\dot{w}(\tau)\} = \sigma\delta(t-\tau)$. Let $Q = \text{diag}(1,0,0,0)$, and $R = [1]$. Use MATLAB to solve the stochastic control problem and find V^*.

Problem 5.5 Consider the inverted-pendulum cart described in problem 2.11 and let $d = \dot{w}$ be a white-noise disturbance with $E\{\dot{w}(t)\dot{w}(\tau)\} = \sigma\delta(t-\tau)$. Solve the stochastic control problem in terms of σ.

Chapter 6

The LQG Problem

In this chapter we consider the problem of output feedback where only a linear combination of the system states is available for feedback and these "measured" signals are corrupted by noise. In contrast to previous chapters we limit our discussion immediately to the linear time-invariant, steady-state problem. There is no general output-feedback theory for nonlinear stochastic systems, and for linear time-varying systems the theory is considerably more involved than for time-invariant systems. We first derive the equations for optimal state estimation (Kalman-Bucy filter) using the stochastic-deterministic dualism developed in Chapter 5. We then demonstrate a separation principle that is used to compute an optimal state-estimate, feedback-control law. The stochastic regulator problem considered in this chapter is commonly called the linear-quadratic-Gaussian (LQG) problem. Finally, we discuss briefly the problem of LQG design with fixed-order compensators.

6.1 Kalman-Bucy Filter

Consider the problem of estimating the state of the stochastic system

$$\dot{x} = Ax + Bu + \xi \tag{6.1}$$

where the control input u can be measured directly, but the state is available only indirectly through the noisy-output measurement

$$y = Cx + \theta \tag{6.2}$$

where $\xi(t)$ and $\theta(t)$ are uncorrelated zero-mean, Gaussian, white-noise random vectors with correlation matrices

$$E\{\xi(t)\xi'(\tau)\} = \Xi\delta(t-\tau); \quad E\{\theta(t)\theta'(\tau)\} = \Theta\delta(t-\tau) \tag{6.3}$$

We refer to $\xi(t)$ and $\theta(t)$ as the **process noise,** and the **measurement noise**, respectively. In the above we do not use the more rigorous notation of stochastic differential equations because the system is linear with additive noise.

The problem is to find a dynamical system that optimally estimates the state of the system $x(t)$ given the above measurements. Let $\hat{x}$ denote the state estimate and $\tilde{x} = x - \hat{x}$ denote the estimation error. The estimation error to be minimized is assumed to be of the form

$$V_f = \lim_{t\to\infty} E\{\tilde{x}'(t)cc'\tilde{x}(t)\} \tag{6.4}$$

where c' is a given fixed-row vector. We would like our estimator to be optimal for any c selected, e.g., $c' = (1, 0, 0, \ldots)$, which would correspond to minimizing the mean-squared error in the first component of the vector error $\tilde{x}$. To simplify the development we assume the following structure for the optimal-estimator system, although it can be shown that the unconstrained solution may always be expressed in this form (see, e.g., [103]).

$$d\hat{x}/dt = A\hat{x} + Bu + K_f(y - C\hat{x}) \tag{6.5}$$

Note that the form of the estimator appears as a "model" of the **certainty-equivalence system,** $\dot{x} = Ax + Bu$, driven by the output-estimation error, $(y - C\hat{x})$. The design parameter will be the matrix K_f, which we will hereafter refer to as the **filter-gain matrix**. If one subtracts (6.5) from (6.1) and substitutes (6.2), one obtains the following equation for the estimation error

$$d\tilde{x}/dt = (A - K_fC)\tilde{x} + \psi \tag{6.6}$$

where ψ is the zero-mean, Gaussian, white-noise process $\psi = \xi - K_f\theta$ with correlation matrix

$$E\{\psi(t)\psi'(\tau)\} = (K_f\Theta K_f' + \Xi)\delta(t-\tau) \tag{6.7}$$

Note that the process ψ is "white" because the linear combination of two white processes is white, and the correlation matrix contains no "product" terms because the processes ξ and θ are assumed to be uncorrelated. From the analysis of Section 5.6, we know that V_f given by (6.4) may be computed from

$$\begin{aligned} V_f &= tr\ Scc'; \\ 0 &= (A - K_f C)S + S(A - K_f C)' + \Xi + K_f \Theta K_f' \end{aligned} \tag{6.8}$$

If we compare (6.8) with the deterministic LQR problem, which can be written (see Section 5.6),

$$\begin{aligned} V_c &= tr\ Px(0)x'(0); \\ 0 &= (A - BK)'P + P(A - BK) + Q + K'RK \end{aligned} \tag{6.9}$$

we may note the following duals:

- Ξ dual of Q, and Θ dual of R
- c, dual of $x(0)$
- A' dual of A, and C' dual of B
- K_f', dual of K
- S, dual of P

But we know from optimal LQR theory that the value of K that minimizes V_c given by (6.9) is equal to $K = R^{-1}B'P$, where P satisfies the ARE

$$0 = A'P + PA + Q - PBR^{-1}B'P \tag{6.10}$$

Thus, by stochastic-deterministic dualism we deduce that the value of K_f that minimizes V_f given by (6.4) is equal to

$$K_f = SC'\Theta^{-1} \tag{6.11}$$

where S satisfies the **filter algebraic Riccati equation (FARE)**

$$0 = AS + SA' + \Xi - SC'\Theta^{-1}CS \tag{6.12}$$

By dualism, a positive-definite solution S of (6.12) that results in a stable matrix $(A - K_f C)$ *exists if* (A', C') *is stabilizable,* (A', E) *is detectable, where* $\Xi = E'E$*, and* Θ *is positive definite.*

Two properties of the optimal estimator are needed in the derivation of a separation principle in the next section. We state them below without proof. Proofs may be found in [113].

- The estimation vector $\hat{x}$ and the estimation-error vector $\tilde{x}$ are uncorrelated, i.e.,

$$E\{\hat{x}(t)\tilde{x}'(t)\} = O$$

- The process $y - C\hat{x}$, commonly referred to as the **innovations process**, is zero-mean white, with correlation matrix $\Theta\delta(t-\tau)$.

Finally, we list below the optimization equations valid for time-varying systems, also without proof. Derivations of the general time-varying Kalman-Bucy filter may be found in advanced texts on the subject, e.g., Anderson and Moore [4] or Kwakernaak and Sivan [113].

$$K_f(t) = S(t)C'\Theta^{-1} \tag{6.13}$$

where $S(t)$ satisfies the Riccati differential equation

$$\dot{S} = AS + SA' + \Xi - SC'\Theta^{-1}CS; \quad S(t_0) = S_0 \tag{6.14}$$

Note that the matrix $S(t)$ represents the covariance matrix for the estimation error $\tilde{x}(t)$, i.e.,

$$S(t) = E\{\tilde{x}(t)\tilde{x}'(t)\}$$

and that the solution of the FARE requires knowledge of the value of error covariance at the initial time t_0. We do not repeat the structure of the optimal filter in the general case because it is the same as in the steady-state case, except that now K_f is a time-varying gain.

Example 6.1 Consider the problem of finding an optimal state estimator for the system

$$\dot{x} = \begin{bmatrix} 0 & 1 \\ 0 & 0 \end{bmatrix} x + \xi; \quad y = [1 \ \ 0]x + \theta$$

with noise matrices for ξ and θ given by

$$\Xi = \begin{bmatrix} 0 & 0 \\ 0 & 1 \end{bmatrix}; \quad \Theta = [\rho^2]$$

Solution: Note first that from the above dynamics, the second component of the state x is the derivative of the first. To focus on the estimation problem, we have omitted the control term u in the system dynamics. Thus, if no measurement noise were present one could reproduce the state by differentiating $y = x_1$. We can explore this case by letting ρ approach 0. If the problem data is substituted into the FARE, (6.12), a solution for S can be computed directly. One obtains

$$S = \begin{bmatrix} \sqrt{2}\rho^{3/2} & \rho \\ \rho & \sqrt{2}\rho^{1/2} \end{bmatrix}$$

with the corresponding filter gain

$$K_f = SC'\Theta^{-1} = \begin{bmatrix} \sqrt{2}\rho^{-1/2} \\ \rho^{-1} \end{bmatrix}$$

The estimator dynamics are given by

$$d\hat{x}/dt = (A - K_f C)\hat{x} + K_f y$$

where K_f is the value computed above. From these dynamics one can compute the transfer function from output y to the optimal estimate $\hat{x}_2$, i.e.,

$$\hat{x}_2/y = \frac{s}{\rho s^2 + \sqrt{2\rho}s + 1}$$

From this transfer function it is clear that as ρ approaches 0, this component of the estimator approaches a differentiator. This is an example of "cheap estimation," the dual of cheap control.

$\triangle$

For single-output systems the optimal filter gain, which may be written as a column vector k_f, may be computed by spectral factorization, using the dual of the theory developed in Section 4.2. In particular, if $\psi(s)$ is the stable spectral factor of the polynomial

$$\phi(-s)\phi(s) + c'M(-s)\Xi M'(s)c \tag{6.15}$$

where $\phi(s)$ and $M(s)$ are defined as in Section 4.2, and where the output matrix C is written as the row vector $C = c'$, then k_f may be computed from the equation

$$\phi(s) + c'M(s)k_f = \psi(s) \tag{6.16}$$

6.2 LQG Solution: Separation Principle

The steady-state LQG problem we will consider is the following: *Find a compensator, possibly dynamic, that uses measurements of* u *and* $y = Cx + \theta$ *to generate a control input* u *that minimizes the performance measure*

$$V = \lim_{t \to \infty} E\{x'Qx + u'Ru\} \tag{6.17}$$

given the stochastic system (6.1) and noise matrices Ξ *and* Θ *for the white-noise processes* ξ *and* θ.

The solution of this problem depends on a separation principle that may be stated as follows:

Separation Principle. *The optimal LQG problem may be solved by separately solving the optimal-estimation problem and the deterministic certainty-equivalence control problem.*

Proof: Note that since $\hat{x}$ and $\tilde{x}$ are uncorrelated

$$\begin{aligned} V &= E\{x'Qx + u'Ru\} \\ &= E\{(\hat{x} + \tilde{x})'Q(\hat{x} + \tilde{x}) + u'Ru\} \\ &= E\{\hat{x}'Q\hat{x} + u'Ru + \tilde{x}'Q\tilde{x}\} \end{aligned} \tag{6.18}$$

Now the dynamic equation for the estimation error, (6.6), depends only on the filter gain K_f *and not on the control input* u.

On the other hand, since the innovation process $y - C\hat{x}$ is zero-mean white noise, the dynamic equation for the estimated state, (6.5), together with the term

$$E\{\hat{x}'Q\hat{x} + u'Ru\}$$

constitute an additive-noise state-feedback problem in $\hat{x}$. By certainty equivalence this problem has an optimal solution

$$u = -K_c\hat{x}, \quad K_c = R^{-1}B'P \tag{6.19}$$

where P satisfies the ARE

$$0 = A'P + PA + Q - PBR^{-1}B'P \tag{6.20}$$

Thus, V given by (6.18) can be minimized by separately minimizing with respect to K_f, which leads to the FARE, given by (6.12),

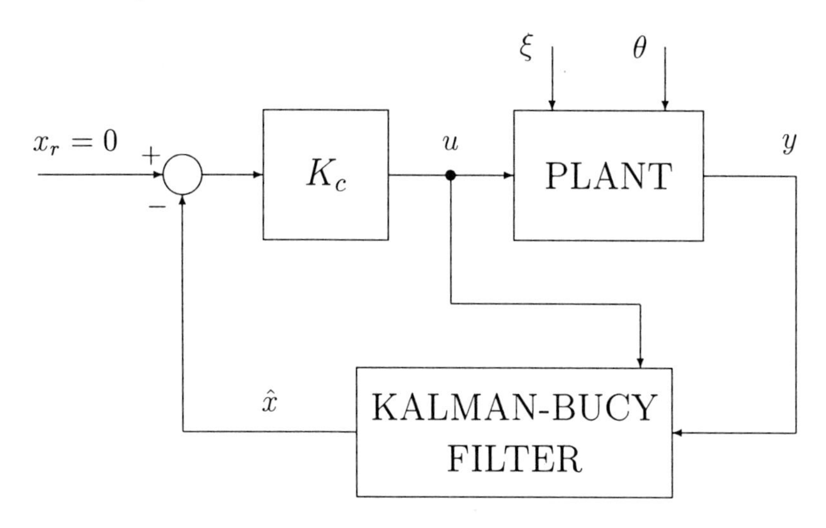

Figure 6.1: State-Estimate Feedback Realization

and using $u = -K_c\hat{x}$, where K_c is computed from the ARE given by (6.20).

The above separation principle demonstrates that the LQG problem can be reduced to the solution of two decoupled Riccati equations, (6.12) and (6.20), and that the optimal compensator is dynamic of order equal to the order of the original plant.

The final LQG controller may be realized in two ways. One way is to separately implement the Kalman-Bucy filter, thus generating $\hat{x}$, and then multiply the output of the Kalman-Bucy filter by $-K_c$ to generate the control input $u = -K_c\hat{x}$. We will refer to this as the **estimator realization**. See Figure 6.1 for the estimator (state-estimate feedback) realization. This approach has the advantage of having a compensator structure that is always stable, since the Kalman-Bucy filter is always stable. It has the disadvantage, however, of requiring measurement of the control-input signal u. Another way to realize a compensator is to compute an equivalent feedback transfer matrix, say $F(s)$, from the output y to the input u. We will refer to this as the **cascade realization**. See Figure 6.2 for a realization of this type. If $u = -K_c\hat{x}$ is substituted back into the state-estimation equations (6.5) one obtains

$$\frac{d\hat{x}}{dt} = (A - BK_c - K_fC)\hat{x} + K_fy \tag{6.21}$$

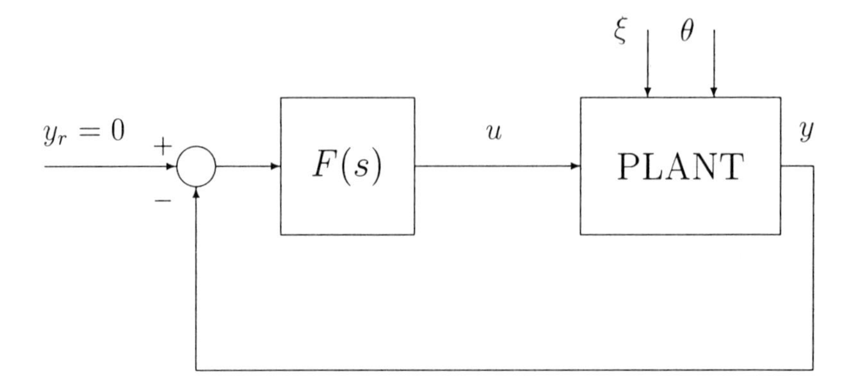

Figure 6.2: Cascade Realization

The transfer function from y to $-u$ can then obviously be written

$$F(s) = K_c(sI - A + BK_c + K_fC)^{-1}K_f \tag{6.22}$$

For convenience, a transfer function realization of the form

$$F(s) = C_f(sI - A_f)^{-1}B_f + D_f \tag{6.23}$$

is commonly written

$$F(s) := \left[\begin{array}{c|c} A_f & B_f \\ \hline C_f & D_f \end{array}\right] \tag{6.24}$$

The state-space cascade realization may thus be written

$$F(s) := \left[\begin{array}{c|c} A - BK_c - K_fC & K_f \\ \hline K_c & O \end{array}\right] \tag{6.25}$$

Note that while $(A - BK_c)$ and $(A - K_fC)$ are both stable matrices, the matrix $(A - BK_c - K_fC)$ is not necessarily stable, so that for some problems, even with a stable plant, the cascade realization will require an unstable compensator. This is a disadvantage of the cascade realization.

The minimal value of the LQG performance measure given by (6.17) may be computed from

$$V^* = tr\{PK_f\Theta K_f'\} + tr\{SQ\} \tag{6.26}$$

where $K_f = SC'\Theta^{-1}$, and where P satisfies the control Riccati equation, (6.10), and S satisfies the filter Riccati equation, (6.12).

Proof: The proof of this result is based on the fact that since $\hat{x}(t)$ and $\tilde{x}(t)$ are uncorrelated, V^* can be expressed as $V^* = V_c + V_f$, where

$$V_c = E\{\hat{x}'Q\hat{x} + u'Ru\} \tag{6.27}$$

and

$$V_f = E\{\tilde{x}'Q\tilde{x}\} \tag{6.28}$$

(see (6.18)). Furthermore, the Markov processes $\hat{x}(t)$ and $\tilde{x}(t)$ satisfy the equations

$$\frac{d\hat{x}}{dt} = A\hat{x} + Bu + K_f(y - C\hat{x}) \tag{6.29}$$

and

$$\frac{d\tilde{x}}{dt} = (A - K_fC)\tilde{x} + \psi \tag{6.30}$$

where $\psi = \xi - K_f\theta$. Recall that $\psi(t)$ is a white-noise process with covariance matrix $K_f\Theta K_f' + \Xi$ and that $K_f(y - C\hat{x})$ is a white-noise process with covariance matrix $K_f\Theta K_f'$. From the additive-noise results in Section 5.3, it then follows that the value of V_c given by (6.27), subject to (6.29) for $\hat{x}(t)$, is given by

$$V_c = tr\{PK_f\Theta K_f'\} \tag{6.31}$$

and that the value of V_f given by (6.28) subject to (6.30) for $\tilde{x}(t)$ is given by

$$V_f = tr\{SQ\} \tag{6.32}$$

and since $V^* = V_c + V_f$, this completes the proof. ■

Example 6.2 Consider the system

$$\dot{x} = \begin{bmatrix} 0 & 1 \\ 0 & 0 \end{bmatrix} x + \begin{bmatrix} 0 \\ 1 \end{bmatrix} u + \xi; \quad y = (1 \;\; 0)x + \theta$$

with performance measure

$$V = \lim_{t\to\infty} E\{x'(t)Qx(t) + u'(t)Ru(t)\}$$

where

$$Q = \begin{bmatrix} 1 & 0 \\ 0 & 0 \end{bmatrix}; \; R = [1]$$

and where the noise matrices for ξ and θ are given by

$$\Xi = \begin{bmatrix} 0 & 0 \\ 0 & 1 \end{bmatrix}; \; \Theta = [1]$$

Solution: The solutions for S and P, from (6.12) and (6.20), respectively, are

$$S = P = \begin{bmatrix} \sqrt{2} & 1 \\ 1 & \sqrt{2} \end{bmatrix}$$

with

$$K_c = R^{-1}B'P = [1 \;\; \sqrt{2}]; \;\; K_f = \begin{bmatrix} \sqrt{2} \\ 1 \end{bmatrix}$$

The cascade compensator realization is then

$$F(s) := \left[\begin{array}{c|c} A_f & B_f \\ \hline C_f & D_f \end{array}\right]$$

where $B_f = K_f$, $C_f = K_c$, $D_f = 0$, and

$$A_f = \begin{bmatrix} -\sqrt{2} & 1 \\ -2 & -\sqrt{2} \end{bmatrix}$$

For this example, we calculate

$$V^* = tr\{PK_f\Theta K_f'\} + tr\{SQ\} = 6\sqrt{2}$$

$\triangle$

Finally we list below, without proof, the solution for the time-varying LQG problem

$$u = -K_c(t)\hat{x}(t); \;\; K_c(t) = R^{-1}B'P(t) \tag{6.33}$$

where $P(t)$ satisfies the Riccati differential equation

$$\begin{aligned} -\dot{P} &= A'P + PA + Q - PBR^{-1}B'P; \\ P(T) &= M \end{aligned} \tag{6.34}$$

and $\hat{x}$ is the optimal state estimate, satisfying (6.21), where $K_f(t) = S(t)C'\Theta^{-1}$ and with $S(t)$ given as the solution of the filter Riccati differential equation, (6.14). The above solution minimizes the performance measure

$$V = E\{\int_t^T (x'Qx + u'Ru)d\tau + x'(T)Mx(T)\} \tag{6.35}$$

6.3 Fixed-Order Compensators

In all of the LQ optimization we have done so far, the structure of the controller has not been constrained a priori. The constraint placed on the structure of the Kalman-Bucy filter was not a real constraint, since it is possible to show that the structure assumed actually corresponds to an unstructured optimal solution. However, there are many practical reasons for placing constraints on the order or structure of the final controller (compensator). For example, it may be necessary to have a simple controller so that controller parameters can be easily tuned online if necessary. Or the compensator may be constrained to be decentralized (diagonal) due to limited measurements. In any case, in this section we briefly outline what can be done with linear-quadratic design for fixed-order, fixed-structure compensators. Unfortunately, the theory in this case is very limited, and one must resort to numerical optimization techniques, such as gradient-search techniques. Gradient methods (see, e.g., [124]) are particularly appropriate for LQ problems because the required gradients are easy to compute. We will use gradients here to optimize with fixed-order, fixed-structure compensators. In particular, consider the following LQG problem: Find a fixed-order, fixed-structure dynamic compensator of the form

$$\begin{aligned} \dot{x}_f &= A_f x_f + B_f y; \\ -u &= C_f x_f \end{aligned} \tag{6.36}$$

that minimizes the performance measure

$$V = \lim_{t\to\infty} E\{x_p'Qx_p + u'Ru\} \tag{6.37}$$

given the system

$$\begin{aligned} \dot{x}_p &= Ax_p + Bu + \xi; \\ y &= Cx_p + \theta \end{aligned} \tag{6.38}$$

where the random signals ξ and θ are the usual zero-mean, Gaussian, white-noise processes with noise matrices Ξ and Θ. Before going any further we make some important observations.

- The order of the compensator is given by $dim\ x_f$, and generally $dim\ x_f < dim\ x_p$. However, if one selects $dim\ x_f$ to be less than $dim\ x_p - dim\ y$, then special care is required to guarantee that even one stabilizing controller, much less an optimal one, exists. Also, since with measurement noise present, the compensator must be **strictly proper**, otherwise white noise will couple directly to the performance measure and V will not be finite. This means that $dim\ x_f \geq 1$ and that we cannot use static-output feedback, at least not for the problem defined above. Of course, if no measurement noise is present this last constraint can be removed.

- The compensator matrices A_f, B_f, and C_f may be of fixed structure and depend on a parameter k. This is what we mean when we say the compensator is of **fixed structure**. The alternative is that all entries in these matrices can be varied independently. This would then be the **fixed-order, free-structure** case.

- All of the usual assumptions will be assumed to hold, e.g., stabilizability of the pair (A, B), positive definiteness of R and Θ, etc., so that if one selects $dim\ x_f = dim\ x_p$ the standard LQG results will be reproduced in the free-structure case.

- It is often not necessary to actually minimize V. In many practical situations it is sufficient to guarantee that V has some specified upper bound, i.e., that V satisfy $V \leq \alpha$.

We now proceed to outline a gradient approach to the above problem. To simplify our discussion we will assume for the moment that the design parameter k is a scalar. First note that the compensator and system (plant) dynamics can be combined to form the composite dynamics

$$\dot{x} = Fx + w \tag{6.39}$$

where

$$F = \begin{bmatrix} A & -BC_f \\ B_f C & A_f \end{bmatrix}; \quad x = \begin{bmatrix} x_p \\ x_f \end{bmatrix} \tag{6.40}$$

and the noise matrix for the augmented white-noise process

$$w = \begin{bmatrix} I & O \\ O & B_f \end{bmatrix} \begin{bmatrix} \xi \\ \theta \end{bmatrix}$$

is given by

$$\tilde{W} = \begin{bmatrix} \Xi & O \\ O & B_f \Theta B_f' \end{bmatrix} \tag{6.41}$$

The performance measure V can also be expressed in terms of the augmented state x as follows

$$V = \lim_{t \to \infty} E\{x' \tilde{Q} x\} \tag{6.42}$$

where

$$\tilde{Q} = \begin{bmatrix} Q & O \\ O & C_f' R C_f \end{bmatrix} \tag{6.43}$$

From the analysis in Section 5.6 we know that V given by (6.42) may be computed from

$$V = tr\ S\tilde{Q}; \quad 0 = FS + SF' + \tilde{W} \tag{6.44}$$

All of the matrices $F, \tilde{W}$, and $\tilde{Q}$ depend on the design parameter k; hence, so does the solution S of the Lyapunov equation $0 = FS + SF' + \tilde{W}$. If we now take the gradient (simple partial derivative in the scalar case) of V with respect to k we obtain

$$\frac{\partial V}{\partial k} = tr\ (\frac{\partial S}{\partial k}\tilde{Q} + S\frac{\partial \tilde{Q}}{\partial k}) \tag{6.45}$$

where $\partial S / \partial k$ may be computed, from (6.44), as the solution of the Lyapunov equation

$$0 = F\frac{\partial S}{\partial k} + \frac{\partial S}{\partial k}F' + [\frac{\partial F}{\partial k}S + S\frac{\partial F'}{\partial k} + \frac{\partial \tilde{W}}{\partial k}] \tag{6.46}$$

Now by using the trick of switching the trace coefficient with the constant term in the Lyapunov equation, in this case $\tilde{Q}$ in (6.45) with the bracketed term in (6.46), we obtain with $U = \frac{\partial S}{\partial k}$

$$\frac{\partial V}{\partial k} = tr\ [U\left(\frac{\partial F}{\partial k}S + S\frac{\partial F'}{\partial k} + \frac{\partial \tilde{W}}{\partial k}\right) + S\frac{\partial \tilde{Q}}{\partial k}] \tag{6.47}$$

where S and U satisfy the "dual" Lyapunov equations

$$0 = FS + SF' + \tilde{W} \tag{6.48}$$

and

$$0 = F'U + UF + \tilde{Q} \tag{6.49}$$

Note that (6.47) allows one to treat S and U in V given by

$$V = tr\,[U(FS + SF' + \tilde{W}) + S\tilde{Q}] \tag{6.50}$$

as matrices that are independent of k, as far as computing gradients is concerned.

We now outline a gradient solution to the fixed-order, fixed-structure LQG problem defined above.

1. Select any k_i so that the augmented matrix $F = F(k_i)$ is stable. *This is a nontrivial step when* $dim\ x_f < dim\ x_p - dim\ y$. In fact, there are no known necessary and sufficient conditions on the system matrices A, B, and C for the stabilizability of F. The best one can do is to use some algebraic stability criterion, such as Routh-Hurwitz, and attempt to test for the satisfaction of the resulting inequalities. See [2] and [6] for a general discussion of the fixed-structure stabilization question.

2. Substitute the value of k_i selected above into the matrices F, $\tilde{W}$, and $\tilde{Q}$ and solve the two Lyapunov equations (6.48) and (6.49). Standard software packages are available to solve Lyapunov equations (see, e.g., [80]).

3. Evaluate $\partial V/\partial k$ from (6.47) and use the gradient update of k_i given by

$$k_{i+1} = k_i - \frac{\partial V}{\partial k}\,\beta$$

where β is a "small" positive number, to obtain an improved value of k.

4. Use V given by (6.50) to compute the value of $V_{i+1} = V(k_{i+1})$. If $V_{i+1} \leq \alpha$, stop. Otherwise, go to step 1 and iterate. Of course, if we need to iterate we are faced with the same difficult stability problem mentioned in step 1. As pointed out in [113],

however, at the stability boundary the performance measure approaches infinity and forms a natural barrier to instability, provided only that the parameter β is "small enough."

Note that if k is one of the entries of the compensator matrices, the matrices $\tilde{W}$ and $\tilde{Q}$ will be quadratic in k, and F will be linear in k. Thus, all the trace terms in (6.47) are either constant or linear. This generalizes easily even if the design parameter k is replaced by a matrix of parameters, e.g., representing $K = A_f$. In this case, one simply replaces $\partial V/\partial k$ by $\partial V/\partial K$, where $\partial V/\partial K$ is a matrix with (i,j) entry given by $\partial V/\partial k_{ij}$, where k_{ij} represents the (i,j) entry of the design-matrix K. The following rules for gradients of trace functions are useful in dealing with the matrix-design case, (see [9] for more details on matrix gradients):

$$\frac{\partial(tr\ AKB)}{\partial K} = A'B'; \quad \frac{\partial(tr\ AKBK')}{\partial K} = A'KB' + AKB \tag{6.51}$$

Note that it is possible to set the gradient $\partial V/\partial K$, where V is given by (6.50), equal to zero to obtain a linear equation in K from (6.45). This constitutes a necessary condition for K to be optimal. The solution to this linear equation can then be substituted back into (6.48) and (6.49) to obtain **two coupled Riccati equations** whose solutions can then be used to compute the optimal value of K. This is the approach taken in [93]. Unfortunately, little can be said theoretically about these types of coupled Riccati equations, so that often one will have to resort to some numerical optimization procedure, such as the gradient procedure outlined above. As expected, the coupled Riccati equations reduce to two decoupled Riccati equations when the compensator is assumed to have the estimator realization and the design matrix K is given by $K = [K_c, K_f]$. We carry out some of the steps of the gradient approach on a simple fixed-order LQG problem in the next example.

Example 6.3 Consider the system

$$\dot{x_p} = \begin{bmatrix} 0 & 1 \\ -1 & -2 \end{bmatrix} x_p + \begin{bmatrix} 0 \\ 1 \end{bmatrix} u + \xi; \quad y = [1 \ \ 0]x_p + \theta$$

with LQG matrices $Q = \Xi = I$ and $R = \Theta = [1]$. The optimal LQG cascade compensator, $F(s)$, for this problem will be second order.

Explore the gradient approach to selecting an optimal value of k if $F(s)$ is assumed to be of first order and of the form

$$F(s) = \frac{k}{s+1}$$

with realization $A_f = -1, B_f = k, C_f = 1$.
Solution: First note that the closed-loop characteristic polynomial with this compensator is given by $s^3 + 3s^2 + 3s + (1+k)$. A Routh-Hurwitz test requires for stability that $-1 < k < 8$. *Thus, in starting the gradient algorithm one must pick a value of k within this range.* The matrices required in the gradient algorithm are

$$F = \begin{bmatrix} 0 & 1 & 0 \\ -1 & -2 & -1 \\ k & 0 & -1 \end{bmatrix}; \quad \tilde{W} = \begin{bmatrix} 1 & 0 & 0 \\ 0 & 1 & 0 \\ 0 & 0 & k^2 \end{bmatrix}; \quad \tilde{Q} = \begin{bmatrix} 1 & 0 & 0 \\ 0 & 1 & 0 \\ 0 & 0 & 1 \end{bmatrix}$$

Note that $\tilde{Q}$ is independent of k so that $\partial\tilde{Q}/\partial k = 0$ for this problem. The next step is to substitute a starting value of k into F, $\tilde{W}$, and $\tilde{Q}$ and solve the Lyapunov equations (6.48) and (6.49) for the matrices S and U. All these values are then substituted in (6.47) to compute the gradient of V and the "improved" parameter value k_{i+1}. Note that at this point one must select the parameter step size β. This is another nontrivial task. Generally, one must experiment with "small" values, keeping in mind two important things. One is that k must not jump the stability boundary; the other is that an acceptable level of performance should be reached in a reasonable number of steps. We will not pursue this particular problem any further here; however, we would like to note that the dependence of V on the design parameter k is generally not convex, so that *we are not assured that the gradient method will approach a global minimum.*

6.4 MATLAB Software

The function **lqg** in the **Robust Control Toolbox** of MATLAB [46] will directly compute a cascade compensator for any steady-state LQG problem with given data: A, B, C, Q, R, Ξ, and Θ. In addition, MATLAB allows for some generalizations of the LQG problem considered here. In particular, the loss function can contain cross-coupling between state and input, i.e., $l(x,u) = x'Qx + 2u'N_c'x + u'Ru$; the noise signals can be correlated, i.e., $E\{\xi(t)\theta'(\tau)\} = N_f\delta(t-\tau)$; and the plant need not be strictly proper, i.e., $y = Cx + Du + \theta$. One must form the matrices W and V defined as

$$W = \begin{bmatrix} Q & N_c \\ N_c' & R \end{bmatrix}; \quad V = \begin{bmatrix} \Xi & N_f \\ N_f' & \Theta \end{bmatrix}$$

then, the command

$$[af, bf, cf, df] = lqg(A, B, C, D, W, V)$$

returns a cascade state-space realization of the optimal LQG compensator.

There is no software package in MATLAB that will solve the finite-time LQG problem. One must use the methods of Section 2.4 for the solution of Riccati differential equations.

The MATLAB **Optimization Toolbox** [79] may be used in the design of fixed-order compensators. In the following we present a design for the helicopter example first presented in Chapter 5.

Example 6.4 Let us consider the helicopter of example 5.3 with dynamics

$$\begin{aligned} \dot{x} &= Ax + Bu + \xi \\ y &= Cx + \theta \end{aligned}$$

where the numerical values of A, B, and C are given in example 5.3. In addition, assume that $Q = C'C$, $R = I_{2\times 2}$, $\Xi = BB'$, and $\Theta = I_{2\times 2}$. The problem is to design a compensator $F(s)$ as in Figure 6.2 such that the performance measure

$$V = \lim_{t\to\infty} E\{x'(t)Qx(t) + u'(t)Ru(t)\}$$

is minimized. The required compensator may be computed from the MATLAB function **lqg**. The state-space description of the compensator is given by $F(s) = [af, bf, cf, df]$, where

$$[af, bf, cf, df] = lqg(A, B, C, D, W, V)$$

with $D = 0$ and

$$W = \begin{bmatrix} Q & 0 \\ 0 & R \end{bmatrix}; \quad V = \begin{bmatrix} \Xi & 0 \\ 0 & \Theta \end{bmatrix}$$

With the numerical values given in example 5.3, we obtain

$$af = \begin{bmatrix} -0.0175 & -0.1436 & 0.3852 & -26.3518 \\ 0.0084 & -17.6863 & -4.0536 & -13.9065 \\ 0.0010 & 0.0018 & -6.7274 & -33.2584 \\ 0 & 0.0031 & 1.0000 & -5.1191 \end{bmatrix}$$

$$bf = \begin{bmatrix} 0.0158 & -0.2405 \\ 9.0660 & -0.1761 \\ 0.0091 & 0.2289 \\ -0.0031 & 0.0893 \end{bmatrix}$$

$$cf = \begin{bmatrix} -0.0033 & 0.0472 & 14.6421 & 60.8894 \\ 0.0171 & -1.0515 & 0.2927 & 3.2469 \end{bmatrix}$$

and $df = 0_{2\times2}$. The minimal value of the performance measure V^* is given by

$$\begin{aligned} V^* &= tr\{PK_f\Theta K_f'\} + tr\{SQ\} \\ K_f &= SC'\Theta^{-1} \end{aligned}$$

where the matrices P and S are solutions to the Riccati and filter Riccati equations (2.45) and (6.12), respectively. P and S may be computed from **lqr** and **lqe** as follows:

$$[K, P] = lqr(A, B, Q, R)$$

and

$$[L, S] = lqe(A, G, C, w, v)$$

where $G = I_{4\times4}$, $Q = C'C$, $R = I_{2\times2}$, $w = \Xi = BB'$, $v = \Theta = I_{2\times2}$. The filter gain is $K_f = L$ and the controller gain is K. For the data given above we obtain

$$P = \begin{bmatrix} 0.0071 & -0.0021 & -0.0102 & -0.0788 \\ -0.0021 & 0.1223 & 0.0099 & -0.1941 \\ -0.0102 & 0.0099 & 41.8284 & 174.2 \\ -0.0788 & -0.1941 & 174.2 & 1120.9 \end{bmatrix}$$

$$S = \begin{bmatrix} 8.3615 & 0.0158 & 0.0187 & -0.0042 \\ 0.0158 & 9.0660 & 0.0091 & -0.0031 \\ 0.0187 & 0.0091 & 0.0250 & 0.0040 \\ -0.0042 & -0.0031 & 0.0040 & 0.0016 \end{bmatrix}$$

with $V^* = V_c + V_f = 28.35 + 14.18 = 42.53$.

6.5 Notes and References

What we now call the Kalman-Bucy filter was first reported in 1961 in Kalman and Bucy [103]. The Kalman-Bucy filter has been applied to many control and signal processing problems, and many books have been written on the subject. See e.g., Anderson and Moore [5] and Maybeck [134]. If the measurement-noise matrix Θ is singular the estimation problem becomes "singular" and the solution we have outlined here is not applicable (we require the inverse of Θ). The problem can sometimes be made nonsingular by differentiating the output y.

An early application of the separation theorem was reported in Gunckel and Franklin [84] in 1963. A rigorous derivation of the separation principle for continuous-time systems appeared in Wonham [210] in 1968. The decade 1960–70 witnessed intensive research activities on the LQG problem, and in 1971 a special issue of the *IEEE Transactions on Automatic Control* was published on the subject, edited by Athans [11]; in particular, see the lead article by Athans [10] on the role and use of LQG theory in control-system design. This special issue contains an extensive bibliography (Mendel and Geseking [140]) of papers published up to 1971.

We have assumed here that the process- and measurement-noise signals where noise signals were white and uncorrelated. It is possible to extend the results to correlated and nonwhite (colored) noise signals. The basic idea is to view the colored signals as the outputs of linear systems driven by white noise. The linear system can be computed using spectral-factorization techniques. Further details on the problem of correlated or colored noise signals in LQG design may be found in sect. 3.6 of Lewis [123].

In this chapter we limited our development to the time-invariant, steady-state case. The general time-varying case is developed in most advanced texts on the subject. See, e.g., Anderson and Moore [4] or Kwakernaak and Sivan [113]. Many applications of LQG theory have been reported at conferences and in journals and books, especially in the aerospace field. The reader is referred, in particular, to the text of McLean [137] for aerospace applications. Further applications of LQG theory may be found in Athans [11], Dressler and Tabak [61], and Gangsaas [73]. Some of the problems in LQG control associated with plant perturbations and noise uncertainty are discussed in Chen and Dong [44]. See also Safonov and Athans [180] and [181] on gain and phase margins of multiloop LQG problems.

We have not discussed at all the frequency-domain (Wiener) approach to LQG problems, exemplified by the text of Newton, Gould, and Kaiser [151] and more recently developed by Youla, Bongiorno, and Jabr [213] for multivariable systems. The frequency-domain approach for multivariable systems requires a considerable amount of mathematics and is limited to time-invariant systems. It does, however, provide an alternative design approach, which may be valuable when data are given in the frequency domain or one does not want to be limited to dynamic compensators (proper compensators). The reader interested in this approach is referred to [213] or recent texts that develop this approach, e.g., Vidyasagar [199]. Frequency-domain LQG theory is commonly referred to as H^2 control since the performance measure in the frequency domain corresponding to the time domain LQG measure is an H^2 norm defined on a stable transfer function $G(s)$, i.e.,

$$\| G(s) \|_2^2 = \int_{-\infty}^{\infty} tr\ G'(-j\omega)G(j\omega)d\omega \tag{6.52}$$

In our development of fixed-order compensators in Section 6.3, we followed the parameter optimization approach in Kwakernaak and Sivan [113]. The text of Vincent and Grantham [200] contains a detailed development of parameter-optimization theory. The work by Grigor'ev and Vorobjov [81] includes some discussion of the complexity associated with the satisfaction of stability inequalities. In Mercadal [142], a homotopy approach is applied to the LQG fixed-order problem. Homotopy methods are based on the parameterization of a complex problem with a scalar parameter α in such a way that with $\alpha = 0$ the problem is reduced to an easily solvable problem and with $\alpha = 1$ the problem is the complex problem at hand. The difficult problem is then approached by small increases in α, from zero to one.

In most practical applications more than just a single performance measure is to be optimized. The theory of **multiobjective optimization**, also referred to as **vector optimization,** is developed in Vincent and Grantham [200]. Applications of vector optimization to LQG problems may be found in Khargonekar and Rotea [107], Mäkilä [129], and Toivonen [192].

The issue of accuracy in simulation of an analog LQG solution on the digital computer is discussed in Gevers and Li [75] and Oranç and Philips [155].

Finally, it should be noted that the gradient approach developed here for the LQG problem can also be used to solve LQR problems when one is constrained to use static-output feedback, i.e., $u = -K_s y$, where $y = Cx$ is the output of the plant, see, e.g., Levine, Johnson, and Athans [120]. In this case, the matrices $F, \tilde{W}$, and $\tilde{Q}$ become

$$F = A - BK_sC; \quad \tilde{W} = x(0)x'(0); \quad \tilde{Q} = Q + C'K_s'RK_sC \qquad (6.53)$$

for the steady-state LQR problem

$$\begin{aligned} \dot{x} &= Ax + Bu \\ y &= Cx \\ V &= \int_0^\infty (x'Qx + u'Ru)dt \end{aligned}$$

The optimal controller for this problem formulation depends on the specific initial state $x(0)$, as can be seen from the expression for $\tilde{W}$ in (6.53); hence, this problem is sometimes (as in Anderson and Moore [3]) referred to as the **specific optimal regulator problem**. To remove the dependency on a specific initial state, one often assumes $x(0)$ is random and averages V over all possible initial states. In particular, if $x(0)$ is zero mean with covariance matrix $\Sigma = E\{x(0)x'(0)\}$, then in the specific regulator problem discussed above we have $\tilde{W} = \Sigma$. This is still a "specific" problem because although we have optimized for a random set of initial states it is still a problem specific to some properties for the initial states, i.e., the covariance Σ. As in the case of fixed-order compensators, a critical problem here is the existence of a static output-feedback law that stabilizes the closed-loop system. It is interesting to note that the existence of stabilizing static-output feedback is equivalent to the solution of a specially formulated optimal LQR problem with a cross-coupled performance measure (see Trofino-Neto and Kučera [195]).

6.6 Problems

Problem 6.1 Consider the system

$$\dot{x} = \begin{bmatrix} 1 & 1 \\ 0 & 1 \end{bmatrix} x + \begin{bmatrix} 0 \\ 1 \end{bmatrix} u + \xi; \quad y = [1 \ \ 0]x + \theta$$

where the noise matrices for ξ and θ are given by

$$\Xi = \sigma \begin{bmatrix} 1 & 1 \\ 1 & 1 \end{bmatrix}; \quad \Theta = [1]$$

Use "dual" spectral factorization to directly compute the optimal filter gain K_f for this problem. Write out the equations required to solve the FARE for this problem. How would you solve this equation?

Problem 6.2 Show that the eigenvalues of the optimal state-estimate feedback system are given by the eigenvalues of $(A - BK_c)$ plus the eigenvalues of $(A - K_f C)$. Hint: Write out the differential equations for $\hat{x}$ and $\tilde{x}$.

Problem 6.3 Consider the system in problem 6.1, with a performance measure

$$V = \lim_{t \to \infty} E\{x'(t)Qx(t) + u^2(t)\}$$

where

$$Q = \rho \begin{bmatrix} 1 & 1 \\ 1 & 1 \end{bmatrix}$$

Compute a realization of the optimal cascade compensator. Compute the eigenvalues of the closed-loop system.

Problem 6.4 Consider the fixed-order compensator example in Section 6.3. Compute the optimal cascade LQG filter for this example (use a computer software package if necessary). Start with a design parameter value of $k = 7$ and do three iterations of the gradient algorithm presented in Section 6.3 (again, use a software package if necessary). Compare the value of performance obtained by the three iterations. Experiment with parameter step sizes, β, of 0.01, and 0.1.

Problem 6.5 Write out all the equations for the specific optimal LQR problem that must be solved if one wants to compute the optimal static-output feedback matrix K_s (see discussion in Section 6.5). Show that these equations reduce to the standard LQR equations when the full state is available, i.e., $C = I$.

Problem 6.6 Figure 6.3 describes a linearized quarter-car model of the active suspension of a car. It first appeared in [198] and was later studied in [170]. The model is described by the following equations:

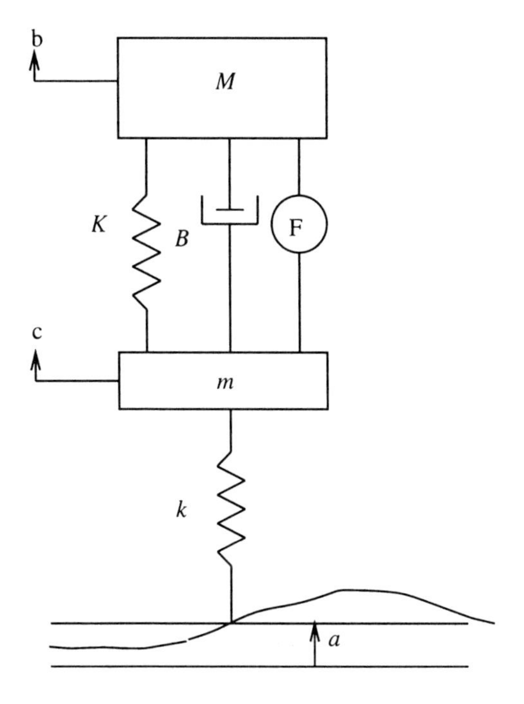

Figure 6.3: Linear Quarter-Car Model

$$
\begin{aligned}
\dot{x}(t) &= Ax(t) + Bu(t) + G\xi(t) \\
y(t) &= Cx(t) + \theta(t)
\end{aligned}
$$

where

$$
\begin{aligned}
A &= \begin{bmatrix} 0 & 1 & 0 & 0 \\ -\frac{k}{m} & -\frac{B}{m} & \frac{K}{m} & \frac{B}{m} \\ 0 & -1 & 0 & 1 \\ 0 & \frac{B}{m} & -\frac{K}{M} & -\frac{B}{M} \end{bmatrix} \\
B &= \begin{bmatrix} 0 \\ \frac{M}{m} \\ 0 \\ -1 \end{bmatrix}; \quad G = \begin{bmatrix} -1 \\ 0 \\ 0 \\ 0 \end{bmatrix} \\
C &= \begin{bmatrix} 0 & 0 & 1 & 0 \end{bmatrix}
\end{aligned}
$$

The system is driven by a zero-mean, Gaussian, white-noise ξ, and the measurement noise θ is also zero-mean, Gaussian white noise, where

$$
E[\xi(t)\xi(\tau)] = \Xi\delta(t-\tau)
$$

m	30 Kg
M	250 Kg
k	150,000 N/m
K	15,000 N/m
B	1,000 $N/m/s$
Ξ	$7 \times 10^{-4}(m/s)^2$
Θ	$10^{-8}m^2$
q_1	5,000
q_3	50,000
ρ	0

Table 6.1: Parameters for Active Suspension System

$$E[\theta(t)\theta(\tau)] = \Theta\delta(t-\tau)$$

Physically, the state-vector elements are tire deflection $x_1 = c - a$, unsprung mass velocity $x_2 = \dot{c}$, suspension stroke $x_3 = b - c$, and sprung mass velocity $x_4 = \dot{b}$. The input $u = F/M$ represents the active control element, and $\xi = \dot{a}$ is a disturbance due to road roughness. The measured output is the suspension stroke $x_3 = b - c$. Let the performance index be given as

$$V = \lim_{t\to\infty} E\{q_1 x_1^2 + q_3 x_3^2 + \dot{x}_4^2 + \rho u^2\}$$

The term $\dot{x}_4$ has been included in the performance index to help minimize the acceleration of the vehicle. Note that the system dynamics may be used to express $\dot{x}_4$ in terms of the state variables and control input.

1. Use computer software to find the optimal LQG compensator $F(s) = [af, bf, cf, df]$, given the data in Table 6.1.

2. Use the compensator obtained in (1) to plot $x_1(t)$, $x_3(t)$, and $u(t)$ for the initial state $x'(0) = [1, 0, 0, 0]$, where $\xi(t) = 0$ and $\theta(t) = 0$.

3. Plot the singular values of $I + F(jw)C[jwI - A]^{-1}B$ for this example. What gain and phase margins do these plots imply? (see problem 4.2).

4. Assume that the maximum control amplitude plotted above saturates the actuator and must be reduced by 10 percent. Explore nonzero values of ρ to achieve this goal. Choose one value of ρ that will satisfy the reduction goal and plot the corresponding $x_1(t), x_2(t), x_3(t)$, and $u(t)$.

Problem 6.7 Consider the LQG problem

$$\begin{aligned} \dot{x} &= \begin{bmatrix} 0 & 1 \\ -3 & -4 \end{bmatrix} x + \begin{bmatrix} 0 \\ 1 \end{bmatrix} u + \begin{bmatrix} 35 \\ -60 \end{bmatrix} \dot{w} \\ y &= [2 \quad 1] x + \theta \end{aligned}$$

where $E\{\dot{w}(t)\dot{w}(\tau)\} = \delta(t-\tau)$, $E\{\theta(t)\theta(\tau)\} = \delta(t-\tau)$, $R = [1]$, $Q = D'D$, with $D = 8.94[5.92 \quad 1]$.

Compute the optimal cascade compensator $F(s)$. Note that $F(s)$ is unstable even though the open-loop system is stable.

Chapter 7

Robustness Design

In all of the previous chapters we have not directly designed for robustness of the closed-loop system. For state-feedback systems the optimal LQR solution had strong robustness properties, but this was only an accidental byproduct of the original problem statement. In the case of stochastic control, one could argue that the stochastic disturbance models are a way of dealing with uncertain system dynamics, and this is certainly true if one can accept a design that is good "on the average." In the case of the optimal LQG solution, we shall see in the first section of this chapter that the solution may actually have arbitrarily small stability margins. We will also see that even the LQR solution, with its strong robustness properties in each input channel, may have arbitrarily small stability margins for general parameter perturbations. Thus, in this chapter we explore some robust-design approaches for linear-quadratic control problems. In particular we present two techniques for robust-system design: the LQG loop-transfer-recovery (LQG/LTR) method and the guaranteed-cost method. We also discuss briefly some H^∞ and linear-matrix-inequality (LMI) methods for robust design.

7.1 The Robustness Problem

Using an example first presented by Doyle [56], we show that the LQG solution may have arbitrarily poor stability margins. In particular, the increasing gain margin at the plant input can be arbitrarily small for certain LQG problems. This is, of course, in sharp contrast to the optimal state-feedback LQR solution, which always has an infinite increasing gain margin.

Consider an LQG problem with the following data:

$$\begin{aligned} A &= \begin{bmatrix} 1 & 1 \\ 0 & 1 \end{bmatrix}; \; B = \begin{bmatrix} 0 \\ 1 \end{bmatrix}; \; C = [1 \;\; 0]; \\ Q &= \rho[1 \;\; 1]'[1 \;\; 1]; \;\; R = [1] \end{aligned} \tag{7.1}$$

and

$$\Xi = \sigma[1 \;\; 1]'[1 \;\; 1]; \;\; \Theta = [1] \tag{7.2}$$

where ρ and σ are positive numbers.

The control gain K_c was computed for this problem in example 4.2 and found to be $K_c = \alpha[1 \;\; 1]$, where $\alpha = 2 + \sqrt{4+\rho}$. The filter gain K_f may be computed, using dualism (see problem 6.1), to be $K_f = \beta[1 \;\; 1]'$, where $\beta = 2 + \sqrt{4+\sigma}$. Let us now assume that the input to the plant experiences a perturbation in gain m so that the plant dynamics are modified to

$$\begin{aligned} \dot{x}_p &= Ax_p + mBu; \\ y &= Cx_p \end{aligned} \tag{7.3}$$

where $m = 1$ corresponds to the nominal plant, and that the filter is given by the usual state-estimate feedback equations

$$\begin{aligned} \dot{x}_f &= (A - K_fC)x_f + K_fy; \\ u &= -K_cx_f \end{aligned} \tag{7.4}$$

where K_c has been computed for the nominal system. Then the state of the closed-loop system satisfies the equation

$$\dot{x} = \begin{bmatrix} A & -mBK_c \\ K_fC & A - BK_c - K_fC \end{bmatrix} x \tag{7.5}$$

where x represents the composite state of plant x_p and filter x_f. With the above numerical values this equation becomes

$$\dot{x} = \begin{bmatrix} 1 & 1 & 0 & 0 \\ 0 & 1 & -m\alpha & -m\alpha \\ \beta & 0 & 1-\beta & 1 \\ \beta & 0 & -\alpha-\beta & 1-\alpha \end{bmatrix} x \tag{7.6}$$

The characteristic polynomial for the system matrix in (7.6) has a constant term (equal to the determinant of the system matrix) equal to $1+(1-m)\alpha\beta$. A necessary condition for the system to be stable is that all the coefficients of the characteristic polynomial be positive. This requires that m satisfy the inequality

$$m < 1 + \frac{1}{\alpha\beta} \tag{7.7}$$

Obviously, from inequality (7.7), the amount of allowable increasing gain may be made arbitrarily close to zero by choosing α and β large enough. It can also be shown (see [56]), that for this example, the stability margin for decreasing gain can be made arbitrarily small.

This example illustrates the fact that the loop-transfer-function matrix for the optimal LQG system, computed at the input point and denoted $H_{LQG}(s)$, may not satisfy Kalman's inequality. In particular, in this case the value of $\underline{\sigma}(I+H_{LQG}(j\omega))$ can be made arbitrarily small for some frequency by choosing α and β large enough.

In the next example, taken from Soroka and Shaked [187], we show that even optimal LQR state feedback may have arbitrarily small stability margins to arbitrary parameter perturbations. Consider the steady-state LQR problem with data

$$A = \begin{bmatrix} -1 & 0 \\ 0 & -2 \end{bmatrix}; \; B = \begin{bmatrix} 1 \\ 1 \end{bmatrix}; \; Q = \begin{bmatrix} 1 & -1 \\ -1 & 1 \end{bmatrix}; \; R = \rho[1]$$

If the input matrix B is perturbed to $B = (1+\epsilon)[1 \;\; 1]'$ we know that closed-loop stability of the optimal system is preserved for any ϵ that satisfies the inequality $\epsilon > -1/2$, since the optimal LQR solution is known to have a decreasing gain margin of at least 1/2 in each channel (see Section 4.3). We will show next, however, that for the perturbation $B_\epsilon = [1+\epsilon \;\; 1]'$ the lower bound on ϵ becomes arbitrarily small as ρ approaches zero; i.e., as one attempts to implement "cheap control." Using spectral factorization it is not difficult to show that the optimal gain for $\epsilon = 0$ is given by

$$K_c = [1 + q - \sqrt{5+2q} \qquad 2\sqrt{5+2q} - q - 4]$$

where $q = \sqrt{4 + 1/\rho}$. Also, by direct computation the characteristic polynomial for the closed-loop matrix $A - B_\epsilon K_c$ is given by $s^2 + d_1 s + d_2$, where

$$d_1 = \sqrt{5+2q} + \epsilon(1 + q - \sqrt{5+q})$$

For stability it is necessary that $d_1 > 0$. This leads to the inequality

$$\epsilon > \frac{\sqrt{5+2q}}{\sqrt{5+2q}-(1+q)}$$

As ρ approaches zero, q approaches infinity, and the right-hand term in the above inequality approaches a very small negative number. Thus, the value of the first component of B can be decreased only by a very small value, compared with the value of $-1/2$, which is possible when both components of B are perturbed by $1+\epsilon$.

7.2 LQG/LTR Design

In this section we present a method for asymptotically recovering the robustness properties of optimal LQR state feedback, when optimal LQG state-estimate feedback is used in the feedback loop. This method was first proposed by Doyle and Stein in 1979 [60] and is now commonly referred to as the **LQG Loop transfer recovery (LQG/LTR)**.

If the feedback loop is broken at the input to the plant, the LQG loop-transfer function (LTF) may be written

$$H_{LQG}(s) = F(s)C(sI-A)^{-1}B \tag{7.8}$$

where $F(s) = K_c(sI - A + BK_c + K_fC)^{-1}K_f$. See $\times$ in Figure 7.1 and equation (6.22). Now introduce fictitious plant noise into the LQG problem so that the plant noise matrix Ξ becomes $\Xi + rBB'$. To avoid some technical complications we assume $dim\ y = dim\ u$. *We will show that as r approaches infinity, the loop transfer function H_{LQG} given by (7.8) approaches the LQR state-feedback LTF*

$$H_{LQR}(s) = K_c(sI-A)^{-1}B \tag{7.9}$$

when $C(sI-A)^{-1}B$ is minimum phase, i.e., has an inverse with no finite poles in the closed right-half s-plane.

Proof: The proof is based on the fact that as r approaches infinity, the optimal filter gain K_f approaches the value

$$K_f = \sqrt{r}BW\Theta^{-1/2} \tag{7.10}$$

where W is an arbitrary orthogonal matrix, i.e., $W'W = I$, and $\Theta^{-1/2}$ denotes the inverse of the symmetric "square-root" of the matrix Θ, i.e., $\Theta^{1/2}\Theta^{1/2} = \Theta$ and $(\Theta^{1/2})^{-1} = \Theta^{-1/2}$. We will prove this

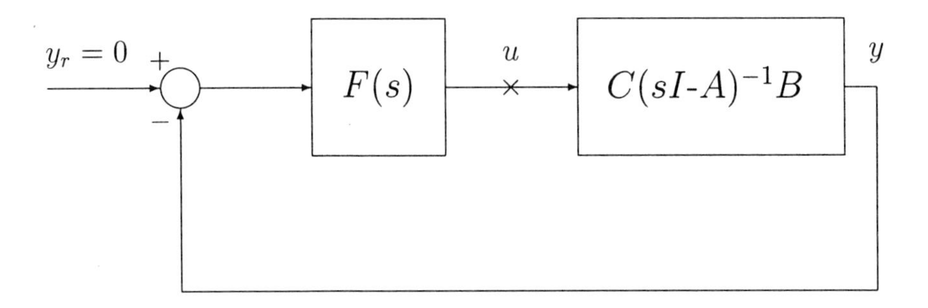

Figure 7.1: Output-Feedback Configuration

later. We show first that if K_f approaches the value given in (7.10), then $H_{LQG}(s)$ given by (7.8) approaches the value of $H_{LQR}(s)$ given by (7.9). To do this we need to make use of the matrix identity $(I+XY)^{-1}X = X(I+YX)^{-1}$, which is easily proven by multiplying on the left by $(I+XY)$ and on the right by $(I+YX)$. Let $\Phi_c(s) = (sI - A + BK_c)^{-1}$, then by elementary matrix manipulations we have

$$\begin{aligned} F(s) &= K_c(sI - A + BK_c + K_fC)^{-1}K_f \\ &= K_c(\Phi_c^{-1}(s) + K_fC)^{-1}K_f \\ &= K_c[\Phi_c^{-1}(s)(I + \Phi_c(s)K_fC)]^{-1}K_f \\ &= K_c(I + \Phi_c(s)K_fC)^{-1}\Phi_c(s)K_f \end{aligned}$$

If the above matrix identity is applied to the last expression for $F(s)$ one obtains

$$F(s) = K_c\Phi_c(s)K_f(I + C\Phi_c(s)K_f)^{-1}$$

If we now use the asymptotic value of K_f given by (7.10) we obtain, using the fact the I is negligible compared with $C\Phi_c(s)K_f$,

$$\begin{aligned} K_f(I + C\Phi_c(s)K_f)^{-1} &= \sqrt{r}BW\Theta^{-1/2}[C\Phi_c(s)\sqrt{r}BW\Theta^{-1/2}]^{-1} \\ &= B[C\Phi_c(s)B]^{-1} \end{aligned}$$

Thus, we have shown that as r approaches infinity, $F(s)$ approaches

$$F(s) = K_c\Phi_c(s)B[C\Phi_c(s)B]^{-1}$$

Let $\Phi(s) = (sI - A)^{-1}$. We show next that

$$\Phi_c(s)B = \Phi(s)B[I + K_c\Phi(s)B]^{-1}$$

To see this note that

$$\begin{aligned}\Phi_c(s)B &= [\Phi^{-1}(s) + BK_c]^{-1}B \\ &= [I + \Phi(s)BK_c]^{-1}\Phi(s)B \\ &= \Phi(s)B[I + K_c\Phi(s)B]^{-1}\end{aligned}$$

In the last step we used once more the matrix identity $(I+XY)^{-1}X = X(I+YX)^{-1}$. Thus, $F(s)$ becomes asymptotically

$$\begin{aligned}F(s) &= K_c\Phi_c(s)B[C\Phi_c(s)B]^{-1} \\ &= K_c\Phi(s)B[I + K_c\Phi(s)B]^{-1}[C\Phi(s)B(I + K_c\Phi(s)B)^{-1}]^{-1} \\ &= K_c\Phi(s)B[C\Phi(s)B]^{-1}\end{aligned}$$

Note that the last equation indicates that as r goes to infinity the equivalent cascade controller $F(s)$ effectively inverts the plant transfer function $C\Phi(s)B$, i.e., $H_{LQG}(s)$ approaches

$$\begin{aligned}H_{LQG}(s) &= F(s)C\Phi(s)B \\ &= [K_c\Phi(s)B(C\Phi(s)B)^{-1}]C\Phi(s)B \\ &= K_c\Phi(s)B \\ &= H_{LQR}(s)\end{aligned}$$

We must verify finally that the asymptotic expression for K_f given by (7.10) is indeed valid. To this end note that the filter gain K_f is given by $K_f = SC'\Theta^{-1}$, where S satisfies the FARE

$$0 = AS + SA' + \Xi + rBB' - SC'\Theta^{-1}CS$$

If we divide this equation by r and take the limit as r goes to infinity, we obtain asymptotically the equation

$$0 = A\frac{S}{r} + \frac{S}{r}A' + BB' - \frac{S}{r}C'\left[\frac{\Theta}{r}\right]^{-1}C\frac{S}{r} \tag{7.11}$$

But (7.11) represents the dual of a cheap-control problem, and since the transfer matrix $C(sI - A)^{-1}B$ is assumed minimum phase it follows that the cheap performance-measure matrix, in this case S/r, approaches zero (see the discussion of cheap control in Sections 2.7 and 4.2). Thus, with $K_f = SC'\Theta^{-1}$ and the factorization $\Theta = \Theta^{1/2}\Theta^{1/2}$, (7.11) becomes

$$0 = BB' - \left(\frac{K_f}{\sqrt{r}}\Theta^{1/2}\right)\left(\frac{K_f}{\sqrt{r}}\Theta^{1/2}\right)' \tag{7.12}$$

Solutions of this equation are of the form

$$\frac{K_f}{\sqrt{r}}\Theta^{1/2} = BW$$

where W is an arbitrary orthogonal matrix. That solutions of this form are solutions of (7.12) is obvious by direct substitution. A proof that this form exhausts all possible solutions of (7.12) may be found in [76], lemma 3.5. From the above equation we obtain the desired asymptotic result

$$K_f = \sqrt{r}BW\Theta^{-1/2}$$

and the proof is complete. ■

The question arises, What happens to LTR when the plant is not minimum phase? We will not pursue this issue in detail but will simply present some results obtained in [214] applicable to the special case of *dim* y = *dim* u and a simple zero in the right-half plane. First, it is shown in [214] that the plant transfer matrix can always be factored as $C(sI - A)^{-1}B = C(sI - A)^{-1}B_mB_z(s)$, where $C(sI - A)^{-1}B_m$ is minimum phase and $B_z(s)$ is "all-pass", i.e., $B_z'(-j\omega)B_z(j\omega) = I$. Let z be a simple zero of the plant in the right-half s-plane, and let w be a nonzero vector such that

$$C(zI - A)^{-1}Bw = 0$$

Such a vector always exists, since the transfer matrix must be singular at a zero. It is then shown in [214] that with the above limiting procedure, the LQG loop gain approaches, in general

$$H_{LQG}(s) = [I + E(s)]^{-1}[H_{LQR}(s) - E(s)] \tag{7.13}$$

where $E(s) = K_c(sI - A)^{-1}(B - B_mB_z(s))$. For a single zero at $s = z$ the expression for $E(s)$ reduces to

$$E(s) = \frac{2z}{s+z}H_{LQR}(z)ww' \tag{7.14}$$

We now make some general observations on the LQG/LTR approach.

- The LQG/LTR approach really involves a tradeoff between robustness and nominal LQG performance, since artificially increasing the system-noise intensity deteriorates the nominal performance of the true-noise problem. Thus, normally the scalar parameter r is increased only enough to give a desired level of robustness. One particular measure of robustness is the smallest singular value of $(I + H_{LQG}(j\omega))$, since the infimum of this over ω relates directly to gain and phase margins (see problem 4.2).

- Since K_c in the above derivation was not necessarily the optimal LQR gain, the LTR method allows one to approach any LTF of the form $K(sI - A)^{-1}B$.

- From (7.14) one can see that if the norm of the matrix $H_{LQG}(z)$ is small, the recovery error will be small even if the plant is nonminimum phase. Stated in engineering terms, good loop recovery is possible even for nonminimum-phase plants if the zero is located sufficiently outside the loop passband.

- By dualism it is easy to see that loop-transfer-function recovery is also possible by artificially increasing the Q matrix to $Q + rCC'$ and keeping the filter gain K_f fixed. In this case, the "observer" loop gain $C(sI - A)^{-1}K_f$ is asymptotically recovered.

- Although the LQG/LTR design approach can be used to guarantee given gain and phase margins, it cannot guarantee stability robustness with respect to arbitrary parameter perturbations.

7.3 Guaranteed-Cost Design

In this section we outline a design procedure proposed by Chang and Peng [41] in 1972 for the design of state-feedback systems that guarantees a certain level of linear-quadratic performance for all admissible plant perturbations. This procedure is now commonly referred to as the **guaranteed-cost** approach. The basic idea is to minimize an upper bound on performance, the upper bound being chosen to be valid for all admissible plant perturbations. Of course, this **robust-performance** solution must also yield a **robustly stable** closed-loop system.

Before proceeding to the design theory, we need some preliminary analysis results on robust stability and robust performance. Consider the perturbed system

$$\dot{x} = (\tilde{A} + \Delta\tilde{A})x \tag{7.15}$$

where $\Delta\tilde{A}$ represents a perturbation, possibly time-varying, of a known constant stable matrix $\tilde{A}$. We first have the following robust stability result:

Given a matrix $Q > 0$ and a symmetric matrix function $\mathcal{U}(\mathcal{P})$ such that for all admissible $\Delta\tilde{A}$ the inequality (quadratic sense)

$$\mathcal{U}(P) \geq (\Delta\tilde{A}'P + P\Delta\tilde{A}) \tag{7.16}$$

holds, then system (7.15) is asymptotically stable for all admissible perturbations if the solution P of the equation

$$0 = \tilde{A}'P + P\tilde{A} + Q + \mathcal{U}(P) \tag{7.17}$$

is positive definite.

Proof: Consider the Lyapunov function $V = x'Px$. Its time derivative for the system (7.15) is

$$\dot{V} = x'(\tilde{A}'P + P\tilde{A} + \Delta\tilde{A}'P + P\Delta\tilde{A})x$$

If the value of $\tilde{A}'P + P\tilde{A}$ from (7.17) is substituted back into this equation, one obtains

$$\dot{V} = -x'Qx - x'[\mathcal{U}(P) - (\Delta\tilde{A}'P + P\Delta\tilde{A})]x$$

From the property of $\mathcal{U}$, i.e., inequality (7.16), and the assumption that $Q > 0$, it then follows that $\dot{V}$ is a negative-definite function. From Lyapunov stability theory, this then means that the system (7.15) is asymptotically stable for all admissible $\Delta\tilde{A}$. ■

We next demonstrate the following robust performance result, taken from [41]. Consider the perturbed system (7.15), with the quadratic performance measure

$$V(x; \Delta\tilde{A}) = \int_0^\infty x'(t)\tilde{Q}x(t)dt, \quad x(0) = x \tag{7.18}$$

If the system (7.15) is asymptotically stable for all admissible perturbations and there exists a constant matrix $\bar{P}$ such that the matrix inequality (quadratic sense) holds

$$0 \geq (\tilde{A} + \Delta\tilde{A})'\bar{P} + \bar{P}(\tilde{A} + \Delta\tilde{A}) + \tilde{Q} \tag{7.19}$$

then

$$V(x; \Delta\tilde{A}) \leq x'\bar{P}x \tag{7.20}$$

for all admissible perturbations.

Proof: Let $x(t)$ denote the solution for the perturbed system (7.15). Multiply (7.19) on the right by $x(t)$ and on the left by the transpose of $x(t)$, and then integrate from $t = 0$ to $t = \infty$. This yields

$$0 \geq \int_0^\infty \frac{d}{dt}[x'(t)\bar{P}x(t)]dt + \int_0^\infty x'(t)\tilde{Q}x(t)dt$$

Now using the fact that $x(\infty) = 0$, since the perturbed system is assumed asymptotically stable, we obtain

$$0 \geq -x'\bar{P}x + V(x; \Delta\tilde{A})$$

where $x = x(0)$. This then establishes the required performance bound.

■

The question arises, How does one find an appropriate bounding $\bar{P}$? *We show next that if $\bar{P} > 0$ satisfies the equation*

$$0 = \tilde{A}'\bar{P} + \bar{P}\tilde{A} + \tilde{Q} + \mathcal{U}(\bar{P}) \tag{7.21}$$

then $\bar{P}$ is the required bounding matrix, i.e., the matrix that satisfies the guaranteed-cost inequality (7.20).

Proof: From the robust stability result derived above it follows, from the fact that $\tilde{Q} > 0$ and $\bar{P} > 0$, that the system (7.15) is asymptotically stable for all admissible perturbations. From (7.21) and the properties of $\mathcal{U}$, inequality (7.19) follows, and from the above robust performance result the proof is complete. ■

A remaining question is, How does one select the matrix function $\mathcal{U}$? The answer depends on the nature of the perturbation $\Delta\tilde{A}$ and how tightly one wants to bound the right-hand term in (7.16). A tradeoff is also required between the complexity of $\mathcal{U}$ and its analytical properties. The two types of perturbations we will consider here are

$$\Delta\tilde{A} = \sum_{0}^{q} w_i \tilde{A}_i \tag{7.22}$$

$$= \sum_{0}^{q} w_i \tilde{D}_i \tilde{E}_i \tag{7.23}$$

where $\tilde{A}_i, \tilde{D}_i$, and $\tilde{E}_i$ are given matrices and w_i are unknown, possibly time-varying, parameters that satisfy the normalized bounding condition $|w_i| \leq 1$. Parameter perturbations of this type are said to be **structured perturbations**, since only selected entries of $\Delta\tilde{A}$ are perturbed, as determined by the matrices $\tilde{A}_i, \tilde{D}_i$, and $\tilde{E}_i$. Note also the similarity of (7.22) with the multiplicative white-noise model discussed in Section 5.4. More will be said about the link to the stochastic model later. For simplicity in the following discussion we assume for the moment that there is only one parameter appearing in $\Delta\tilde{A}$. It is a simple matter to extend the results to multiple parameters. We now explore three bounding functions that have been proposed in the literature for $\mathcal{U}$. The first one we consider appeared in the original paper of Chang and Peng [41]. Also, for convenience we will replace $\bar{P}$ by P in all our future equations. Since the matrix

$$\Delta\tilde{A}'P + P\Delta\tilde{A} = w_1(\tilde{A}_1'P + P\tilde{A}_1) \tag{7.24}$$

is symmetric, there always exists an orthogonal matrix S_1 such that

$$S_1'(\tilde{A}_1'P + P\tilde{A}_1)S_1 = \Lambda_1 \tag{7.25}$$

where Λ_1 is a diagonal matrix. Let $|\Lambda_1|$ denote a matrix where all the diagonal elements of Λ_1 are replaced by their absolute magnitudes, then a bounding function $\mathcal{U}$ is given by

$$\mathcal{U}(P) = S_1|\Lambda_1|S_1' \tag{7.26}$$

We will refer to this as the **Chang-Peng bound**. Note that the dependence of $\mathcal{U}$ on P is implicit through the matrices S_1 and Λ_1.

The next bounding function we consider is developed in Bernstein [22].

$$\mathcal{U}(P) = \frac{1}{\gamma_1}P + \gamma_1 \tilde{A}_1' P \tilde{A}_1 \tag{7.27}$$

where γ_1 is an arbitrary positive constant. This function follows directly from the fact that the inequality

$$(\sqrt{\gamma_1}\tilde{A}_1 - \frac{w_1}{\sqrt{\gamma_1}}I)'P(\sqrt{\gamma_1}\tilde{A}_1 - \frac{w_1}{\sqrt{\gamma_1}}I) \geq 0$$

is true for any positive γ_1, and that $w_1^2 \leq 1$. We refer to this as the **linear bound**.

The last bounding function we consider appears in Petersen and Hollot [166]. The function is

$$\mathcal{U}(P) = P\tilde{D}_1\tilde{D}_1'P + \tilde{E}_1'\tilde{E}_1 \tag{7.28}$$

This function follows from the inequality

$$(w_1\tilde{E}_1 - \tilde{D}_1'P)'(w_1\tilde{E}_1 - \tilde{D}_1'P) \geq 0$$

and the fact that $w_1^2 \leq 1$. We refer to this last bounding function as the **quadratic bound**.

We can now discuss the design problem. Consider the perturbed system

$$\dot{x} = (A + \Delta A)x + (B + \Delta B)u \tag{7.29}$$

Assume the state of the system is available for feedback. Let $u = -Kx$. The **guaranteed-cost design problem** is then to *find the state-feedback gain K such that a perturbation-independent upper bound on the performance measure is minimized.* In particular, an upper bound on the quadratic performance

$$V(x : \Delta A, \Delta B) = \int_0^\infty (x'(t)Qx(t) + u'(t)Ru(t))dt, \quad x(0) = x \tag{7.30}$$

is minimized. Note that in this case the closed-loop matrices $\tilde{A}$ and $\Delta\tilde{A}$, which appear in (7.15), are given by

$$\begin{aligned}\tilde{A} &= A - BK \\ \Delta\tilde{A} &= \Delta A - \Delta BK\end{aligned}$$

and that

$$\tilde{Q} = Q + K'RK$$

Thus, all the matrices that appear in (7.21), including $\mathcal{U}$, are functions of the design matrix K. We use the following performance measure to minimize

$$V = tr\ P \tag{7.31}$$

where the matrix P is a solution to the equation

$$0 = \tilde{A}'P + P\tilde{A} + \tilde{Q} + \mathcal{U}(P) \tag{7.32}$$

This is equivalent to minimizing $x'Px$ for randomized initial states x, with $E\{xx'\} = I$. This in turn is a nonlinear programming problem with equality constraints, given by (7.32). One may use Lagrangian multipliers, in this case represented by the matrix $\mathcal{P}$, to convert this to an unconstrained minimization problem in the Lagrangian function $\mathcal{L}$, i.e.,

$$\mathcal{L} = tr(P + [\tilde{A}'P + P\tilde{A} + \tilde{Q} + \mathcal{U}(P)]\mathcal{P}) \tag{7.33}$$

where now P and K may be viewed as independent optimization variables. Necessary conditions for optimality (see, e.g., [124]), are then

$$\partial\mathcal{L}/\partial K = 0 \tag{7.34}$$
$$\partial\mathcal{L}/\partial P = 0 \tag{7.35}$$
$$\partial\mathcal{L}/\partial \mathcal{P} = 0 \tag{7.36}$$

At this point we make a simplifying assumption to reduce the complexity of the resulting optimization equations. We assume that only the matrix A is perturbed, so that $\tilde{A} = A - BK$, but $\Delta\tilde{A} = \Delta A$. With this assumption $\mathcal{U}(P)$ is independent of K, and (7.34) becomes, using the gradient formulas in (6.51),

$$(-2RK + 2B'P)\mathcal{P} = 0$$

which for a nonsingular $\mathcal{P}$ can be solved for K, i.e.,

$$K = R^{-1}B'P \tag{7.37}$$

where P satisfies the **guaranteed-cost optimization equation**

$$0 = A'P + PA + Q - PBR^{-1}B'P + \mathcal{U}(P) \tag{7.38}$$

which follows from (7.36). The control

$$u = -R^{-1}B'Px \tag{7.39}$$

then yields a closed-loop system that guarantees a minimal bound on the performance measure

$$E\{\int_0^\infty [x'(t)Qx(t) + u'(t)Ru(t)]dt\} \tag{7.40}$$

where the expectation in (7.40) is taken with respect to initial conditions $x(0)$, when $x(t)$ satisfies the equation

$$\dot{x} = (A + \Delta A)x + Bu \tag{7.41}$$

Equation (7.35) yields an equation for the Lagrange-multiplier matrix $\mathcal{P}$, which does not affect the above optimization equation in P but is required to verify that $\mathcal{P}$ is indeed nonsingular. We will not do this here, but with the above assumptions $\mathcal{P}$ is nonsingular. The guaranteed-cost optimization equation specializes to different forms depending on the particular bounding functions $\mathcal{U}$ selected. We discuss briefly each of the three forms cited above, for the general case of q parameters as in (7.22) and (7.23). The generalization from one parameter to q is obtained simply by forming $\mathcal{U}$ as the sum of each individual bounding term.

Chang-Peng Bound. In this case, the optimization equation becomes

$$0 = A'P + PA + Q - PBR^{-1}B'P + \sum_{i=1}^{q} S_i|\Lambda_i|S_i' \tag{7.42}$$

In [41], an algorithm is given for the solution of this equation; however, because of the complicated nonlinear nature of the bounding function $\mathcal{U}$, no proof is available for the convergence of the algorithm. *If, however, a positive-definite solution for P can be found, then one is guaranteed both robust stability and robust performance.*

Linear Bound. If we let $2\alpha = \sum_{i=1}^{q} 1/\gamma_i$, the optimization equation can be written in this case

$$0 = (A + \alpha I)'P + P(A + \alpha I) + Q - PBR^{-1}B'P + \sum_{i=1}^{q} \gamma_i A_i' P A_i \tag{7.43}$$

Note that this optimization equation is the same as the optimization equation for state-dependent noise, i.e., equation (5.36), and

$W = \text{diag}(\gamma_i)$; combined with a "degree of stability" equal to $-\alpha$ (see (2.64). This interesting link between the stochastic uncertain model presented in Chapter 5 and the "deterministic" uncertain model presented in this section is discussed in some detail in [22]. Because the two optimization equations are the same, one may use the computational techniques discussed in Chapter 5 for the solution of the linear-bound optimization equation. Equations of the form (7.43) are sometimes referred to as **generalized algebraic Riccati equations (GARE).** Algorithms for the solution of GARE are given in [188] and [209]. The algorithms are shown to converge if the pair (A, B) is stabilizable and there exists a K such that

$$\| \int_0^\infty e^{(A-BK)'t} \mathcal{U}(I) e^{(A-BK)t} dt \| < 1 \tag{7.44}$$

where

$$\mathcal{U}(I) = 2\alpha I + \sum_{i=1}^{q} \gamma_i A_i' A_i$$

and α is as previously defined. The parameters γ_i are positive but otherwise arbitrary. They may be used to optimize the choice of $\mathcal{U}$.

Quadratic Bound. Since for a given A_i the factors D_i and E_i are not unique, one can write $A_i = (\epsilon_i D_i)(1/\epsilon_i E_i)$. The optimization equation for the quadratic-bound case can then be written

$$0 = A'P + PA + Q + \sum_{i=1}^{q} (1/\epsilon_i^2) E_i' E_i - P(BR^{-1}B' - \sum_{i=1}^{q} \epsilon_i^2 D_i D_i')P \tag{7.45}$$

It is interesting to note that this optimization equation is the same type of nonstandard Riccati equation as that encountered in Section 3.5, i.e., (3.36), in connection with the game-theoretic approach to disturbance rejection. The software discussed in Chapter 3 may be used to solve equations of this form. The parameters ϵ_i may be used to help optimize the choice of P.

Example 7.1 Consider the scalar guaranteed-cost problem

$$\dot{x} = w_1 x + u; \quad V(x) = \int_0^\infty (qx^2 + u^2) dt; \quad x(0) = x; \quad |w_1| \leq 1$$

For this problem the data matrices are:

$$A = [0]; \; A_1 = [1]; \; B = [1]; \; Q = [q]; \; R = [1]$$

and we take the nominal to correspond to $w_1 = 0$.

Nominal optimal solution: We consider first, for comparison purposes, the solution of this problem when the unknown parameter w_1 assumes its nominal value of zero. In this case, $u^* = -Px$, where P satisfies the equation

$$0 = q - P^2$$

The positive solution of this equation is obviously $P = \sqrt{q}$, assuming as usual that $q > 0$. The control law is then $u^* = -\sqrt{q}x$, and the nominal optimal performance value is $V^* = \sqrt{q}x^2$. Finally, the closed-loop dynamics are given by

$$\dot{x} = -(\sqrt{q} - w_1)x$$

Note that the closed-loop system is stable for a constant w_1 that varies symmetrically about zero only if $|w_1| < \sqrt{q}$, and if q is very small the stability margin for the parameter w_1 may be very small; in particular, much less than one.

Guaranteed-cost solution, Chang-Peng bound: For this scalar problem we can select $S_1 = 1$ and $\Lambda_1 = 2P$. This results in $\mathcal{U}(P) = 2P$. Then the guaranteed-cost control is given by $u = -Px$, where P satisfies the equation

$$0 = q - P^2 + 2P$$

which has the positive solution $P = 1 + \sqrt{1+q}$. The cost V is bounded by

$$V(x; w_1) \leq (1 + \sqrt{1+q})x^2$$

for all admissible w_1, i.e., all w_1 that satisfy $|w_1| \leq 1$. The closed-loop dynamics are given by

$$\dot{x} = -[(1 + \sqrt{1+q}) - w_1]x$$

For constant w_1, the closed-loop system is stable if and only if $|w_1| \leq (1 + \sqrt{1+q})$, which is within the design range $|w_1| \leq 1$.

Guaranteed-cost control, Linear bound: In this case, $u = -Px$, where P satisfies the equation

$$0 = q - P^2 + (\gamma_1 + 1/\gamma_1)P$$

The positive solution of this equation is given by

$$P = 1/2(\gamma_1 + 1/\gamma_1) + \sqrt{1/2(\gamma_1 + 1/\gamma_1) + q}$$

If γ_1 is selected so as to minimize P, then $\gamma_1 = 1$ and the solution for the linear bound is the same, in this scalar case, as the solution for the Chang-Peng bound.

Guaranteed-cost solution, Quadratic bound: If we let $D_1 = E_1 = 1$, the optimization equation becomes in this case

$$0 = (q + 1/\epsilon_1^2) - (1 - \epsilon_1^2)P^2$$

The positive solution of this equation

$$P = \sqrt{\frac{q + 1/\epsilon_1^2}{1 - \epsilon_1^2}}$$

requires that $|\epsilon_1| < 1$. Note that with this constraint on ϵ_1 the value of P is always greater than one, so that again robust stability is guaranteed for all admissible w_1. This follows directly from the fact that the closed-loop dynamics can be written $\dot{x} = -(P - w_1)x$.

$\triangle$

It should be noted that in all of the theory developed above, the perturbations $\Delta\tilde{A}$, or equivalently w_i, were not restricted to being time-invariant. This implies a certain conservatism in the stability and robustness bounds when the perturbations are actually time-invariant. However, if the parameters w_i do vary with time, all the cost bounds developed above remain valid.

7.4 A Brief Survey of H^∞ Methods

For **unstructured perturbations** in the frequency domain, H^∞ methods provide the best design tools for robust linear-quadratic design (see, e.g., [199]). However, a development of all the relevant H^∞ theory is beyond the scope of this introductory book. Thus, in this section we simply survey some H^∞ results that have been applied to the linear-quadratic problem, with only limited discussion of the underlying theory.

We show how robust stabilization for unstructured perturbations in the frequency domain is related to H^∞ theory and then outline two techniques that allow for the design of systems that are **robustly stable and nominally optimal**. First, the definition of the H^∞ norm is given.

The H^∞ norm of a stable (proper matrix analytic in Re $s \geq 0$) matrix $H(s)$ is defined as

$$\|H(s)\|_\infty = \sup_\omega \bar{\sigma}(H(j\omega)) \tag{7.46}$$

where $\bar{\sigma}(M)$ denotes the largest singular value of the complex matrix M. Note that from the definition of singular values $\bar{\sigma}(M) = \|M\|$, where $\|M\|$ denotes the regular Euclidean norm of the complex matrix M.

We present next a robust-stability result for multiplicative perturbations at the input to the plant $P(s) = C(sI - A)^{-1}B$. In particular, let $P(s)$ be perturbed to $P(s)(I + R(s))$, where a frequency bound is assumed given on the perturbation matrix $R(s)$, i.e.,

$$\bar{\sigma}(R(j\omega)) < |r(j\omega)|, \quad \text{for all } \omega \tag{7.47}$$

Perturbations of the type described above are referred to as unstructured perturbations since all the entries of the perturbation matrix $R(s)$ are variable; i.e., the matrix $R(s)$ has no special structure. We then have the following result:

The closed-loop system, see Figure 7.1, with compensator $F(s)$ is robustly stabilizing for all multiplicative-input perturbations that satisfy (7.47) and that introduce no new poles in the right-half s-plane, if and only if

$$\|(I + F(s)P(s))^{-1}F(s)P(s)r(s)\|_\infty \leq 1 \tag{7.48}$$

where $r(s)$ is any stable rational function for which $r(s)P(s)$ remains proper, has no finite poles or zeros in Re $s \geq 0$, and satisfies (7.47); and $F(s)$ is nominally stabilizing.

Proof: A complete proof of this result may be found in [199], sect. 7.4. We will simply motivate the sufficiency part of the proof. As noted in Section 4.1, the stability of the closed-loop system described above is determined by the number of encirclements of the origin of the determinant of

$$I + F(j\omega)P(j\omega)(I + R(j\omega)) \tag{7.49}$$

But the matrix in (7.49) can be written

$$(I + FP)(I + (I + FP)^{-1}FPR) \tag{7.50}$$

where for convenience we have suppressed the arguments of the matrix functions. If the nominal closed-loop system is stable, then the term $(I + FP)$ has the correct number of encirclements. If condition (7.48) is satisfied the term $(I + FP)^{-1}FPR$ has a maximal singular value that is less than one at all frequencies, so that the term $(I + (I + FP)^{-1}FPR)$ can never be singular, hence, can never have a zero determinant. This means finally that the correct number of encirclements is preserved for the matrix in (7.49). This is essentially a "small-gain" result of the type already used in Section 4.1.

Of course, the fact that the nominal and perturbed systems must have the same number of unstable poles is critical in the above proof, which is based on the perturbed system having the same number of encirclements of the origin as the nominal system. ■

The robust stability condition (7.48) may also be written

$$|r(j\omega)| \leq \underline{\sigma}(I + [F(j\omega)P(j\omega)]^{-1}), \quad \text{for all } \omega \tag{7.51}$$

This follows from $(I + FP)^{-1}FP = (I + (FP)^{-1})^{-1}$ and $\bar{\sigma}(M^{-1}) = \underline{\sigma}(M)$, where M is any complex matrix. Note that if $F(s) = K(sI - A)^{-1}B$, where K is the optimal LQR feedback gain, then we have

$$\underline{\sigma}(I + H_{LQR}(j\omega)) \geq 1/2 \tag{7.52}$$

where $H_{LQR} = FP$. *This means that the optimal LQR state-feedback solution is robustly stable for all input multiplicative perturbations bounds $r(j\omega)$ that satisfy $|r(j\omega)| \leq 1/2$.* The inequality (7.52) follows from the identity $(I + H^{-1})^{-1} = I - (I + H)^{-1}$ and the inequalities

$$\|(I + H^{-1})^{-1}\| \leq 1 + \|(I + H)^{-1}\|$$

$$\frac{1}{\underline{\sigma}(I + H^{-1})} \leq 1 + \frac{1}{\underline{\sigma}(I + H)} \leq 2$$

The last inequality is a result of the fact that for the optimal $H_{LQR}(s)$ one has $\underline{\sigma}(I+H_{LQR}(j\omega)) \geq 1$, as shown in Section 4.2. Of course, for optimal LQG control one cannot guarantee this level of robust stabilization, except asymptotically with LTR design. One can, however, guarantee robustness for $|r(j\omega)| < \alpha$, where $\alpha < 1/2$, by satisfying the H^∞ bound

$$\|(I + H_{LQG})^{-1} H_{LQG}\alpha\|_\infty \leq 1 \tag{7.53}$$

where $H_{LQG} = FP$ and F is the cascade optimal LQG compensator.

We will use this result shortly in presenting the U-parameter design approach. Next, however, we develop some H^∞ and H^2 results that will also be needed in the sequel. The first result links the satisfaction of an H^∞ norm bound with the solution of a Riccati equation. In particular we have:

If the solution of the (nonstandard) Riccati equation

$$0 = A'P + PA + C'C + PBB'P \tag{7.54}$$

is positive definite, and the pair (A, C) is observable, then the following H^∞ norm bound is satisfied

$$\|C(sI - A)^{-1}B\|_\infty \leq 1 \tag{7.55}$$

Proof: With the Lyapunov function $V = x'Px$ and the fact that $P > 0$ and that the pair (A, C) is observable, it follows that A must be stable, so that the H^∞ is well defined. To prove (7.55) we show, what is equivalent, that $I - G^*G \geq 0$, where $G = C(j\omega I - A)^{-1}B$, and where G^* denotes the complex conjugate transpose of G. Let $\Phi(s) = (sI - A)^{-1}$, then from (7.54) we have

$$\begin{aligned} -A'P - PA &= C'C + PBB'P & (7.56)\\ (-sI - A)'P + P(sI - A) &= C'C + PBB'P & (7.57)\\ B'P\Phi(s)B + B'\Phi'(-s)PB &= B'\Phi'(-s)PBB'P\Phi(s)B \\ &\quad + G'(-s)G(s) & (7.58)\\ I - G'(-s)G(s) &= M'(-s)M(s) & (7.59) \end{aligned}$$

where $M(s) = (I - B'P\Phi(s)B)$. Equation (7.58) is obtained by multiplying (7.57) on the right by $\Phi(s)B$ and on the left by $B'\Phi'(-s)$.

Equation (7.59) is obtained by adding I to both sides of (7.58), transposing appropriate terms across the equality sign and factoring the "right-hand side" of $I - G'(-s)G(s)$. Since $M'(j\omega)M(j\omega) \geq 0$, (7.59) completes the proof. ■

The next result links the H^2 norm of a stable transfer matrix to the solution of a Lyapunov equation. The H^2 norm of a transfer matrix $H(s)$, denoted $\|H(s)\|_2$, is defined as

$$\|H(s)\|_2 = \left(\frac{1}{2\pi}\int_{-\infty}^{\infty} tr[H'(-j\omega)H(j\omega)]d\omega\right)^{1/2} \tag{7.60}$$

Let $H(s)$ have a realization $H(s) = C(sI - A)^{-1}B$ with an impulse-response matrix denoted $h(t) = Ce^{At}B$. We then have the following result:

If the matrix A is stable

$$\|H(s)\|_2 = (tr[QC'C])^{1/2} \tag{7.61}$$

where Q satisfies the Lyapunov equation

$$0 = AQ + QA' + BB' \tag{7.62}$$

Proof: From Plancherel's (Parseval's) theorem, the H^2 norm given in (7.60) may also be written

$$\|H(s)\|_2 = \left(\int_0^{\infty} tr[h'(t)h(t)]dt\right)^{1/2} \tag{7.63}$$

$$= \left(\int_0^{\infty} tr[B'e^{A't}C'Ce^{At}B]dt\right)^{1/2} \tag{7.64}$$

$$= (tr\ PBB')^{1/2} \tag{7.65}$$

where P satisfies the Lyapunov equation

$$0 = A'P + PA + C'C \tag{7.66}$$

From the stochastic-deterministic dualism developed in Section 5.6, we know that the trace function in (7.65) can also be written $tr\ QC'C$, where Q is given by (7.62). This completes the proof. ■

Note finally that if $y = Cx$ and x satisfies the stochastic differential equation $\dot{x} = Ax + B\eta$, where η is a white-noise vector with unit-variance matrix, then we have

$$\lim_{t \to \infty} E\{y'(t)y(t)\} = 1/2\pi \int_{-\infty}^{\infty} tr[H'(-j\omega)H(j\omega)]d\omega \tag{7.67}$$

This completes the link between H^2 norms, LQG performance measures, and mean-squared performance measures.

We now outline some H^∞ design techniques that guarantee robust stability and nominal linear-quadratic performance.

- **Q-parameter/convex-programming design.** It is known (see, e.g., sect. 5.2 of [199]) that all compensators that preserve nominal internal stability can be parameterized in terms of an arbitrary stable matrix $Q(s)$. We will not go into the theory for Q-parameterization here; details may be found in [199]. We will, however, briefly outline, following [31], how this theory can be used to design systems that are nominally optimal and robustly stable.

  A closed-loop system of the type shown in Figure 7.1 is **internally stable** if all of the transfer matrices, $(I + FP)^{-1}, (I + FP)^{-1}F, P(I + FP)^{-1}$, and $P(I + FP)^{-1}F$, are stable. If the plant transfer matrix is written in the matrix-fraction form $P(s) = N(s)D^{-1}(s) = \tilde{D}^{-1}(s)\tilde{N}(s)$, where (N, D) and $(\tilde{N}, \tilde{D})$ are right and left coprime factorizations of $P(s)$, respectively, then all compensators that preserve internal stability of the closed-loop system shown in Figure 7.1 may be parameterized as follows

$$F(s) = [Y(s) - Q(s)\tilde{N}(s)]^{-1}[X(s) + Q(s)\tilde{D}(s)] \tag{7.68}$$

  where $X(s)$ and $Y(s)$ are H^∞ matrices that satisfy the Bezout identity $X(s)N(s) + Y(s)D(s) = I$, and $Q(s)$ is any stable matrix. Details on coprime factorization and the Bezout identity may be found in [199]. This parameterization is referred to as Q-parameterization or sometimes Youla parameterization. It can further be shown that with this parameterization key transfer matrices can be expressed as **affine** functions in the matrix Q; i.e., they can be expressed as constant-plus-linear functions of the form $T_1 + T_2QT_3$. Thus, for example, with the compensator parameterized as in (7.68), the transfer matrices $S = (I + FP)^{-1}$, referred to as the **sensitivity matrix**, and $T = (I + FP)^{-1}FP$, referred to as the **complementary sensitivity matrix**, can be written

$$\begin{aligned} S &= DY - DQ\tilde{N}; \\ T &= (I - DY) + DQ\tilde{N} \end{aligned} \tag{7.69}$$

which are both of the form $T_1 + T_2QT_3$. Software for the computation of T_1, T_2, and T_3 is discussed in Section 7.6. The results in (7.69) follow by substitution of $F(s)$ given in (7.68) into the expressions for S and T and the Bezout identity. Note that a bound on the H^∞ norm of T may be used to guarantee robust stability with respect to multiplicative disturbances. For example, if $\|T(s)\|_\infty \leq 1/\alpha$, then the closed-loop system is guaranteed to be stable for all multiplicative perturbations that satisfy $\bar{\sigma}(R(j\omega)) < \alpha$. On the other hand, the H^2 norm of S may represent the root-mean-square (RMS) tracking error. Thus, one possible linear-quadratic optimization problem, subject to a robust-stability constraint, could be stated as follows:

Minimize

$$\|W_1(s)S(s)\|_2$$

with respect to $Q(s)$, subject to the constraint

$$\|T(s)\|_\infty \leq \gamma$$

Since any norm on a function that is affine in Q is **convex** in Q, the above optimization problem reduces to a convex optimization problem, which always has a unique minimal value. Thus, any minimal search techniques that converge must converge to a **global** minimal. Recall (see [124]), that a subset $\mathcal{S}$ of a vector space is convex if for any x_1 and x_2 belonging to $\mathcal{S}$ the linear combination $\lambda x_1 + (1 - \lambda)x_2$, where $0 \leq \lambda \leq 1$ also belongs to $\mathcal{S}$. Also recall that a scalar function $f(x)$ is convex over a convex domain $\mathcal{S}$ if the inequality

$$f[\lambda x_1 + (1 - \lambda)x_2] \leq \lambda f(x_1) + (1 - \lambda)f(x_2) \tag{7.70}$$

holds for $0 \leq \lambda \leq 1$ and all x_1, x_2 belonging to $\mathcal{S}$ (see [33] for further details on convexity and algorithms for convex programming). Since $Q(s)$ is in an infinite-dimensional space, however, further parameterization is required to obtain numerical solutions. The following parameterization preserves affineness in the vector q

$$Q(s) = \sum_{i=1}^{N} q_i Q_i(s) \tag{7.71}$$

where the $Q_i(s)$ are fixed H^∞ matrices and the q_i are components of the vector q. The steps required in the Q-parameter/convex programming design approach are then the following:

1. Find the matrices $N, D, X, ...$ etc. Explicit state-space formulas are given in sect. 4.2 of [199] for the computation of these matrices.
2. Select the matrices $Q_i(s)$ and a value of N for the reduction of the convex-programming problem to a finite-dimensional problem. This is a difficult step, since the choice of these parameters is not obvious. Also, this step introduces an approximation to the true optimal, and finding a feasible solution of the given form may be a problem.
3. Use some convex-programming algorithm, such as an **ellipsoid algorithm** or a **cutting-plane algorithm** (see [33] for a discussion of convex-programming algorithms) to solve the constrained-optimization problem in the finite-dimensional vector q.

- **U-Parameter design.** From H^∞ theory it is known that compensators $F(s)$ that guarantee the satisfaction of the H^∞ norm in (7.53) can be parameterized by an arbitrary bounded real matrix $U(s)$ (a matrix $U(s)$ whose entries are all H^∞ functions and whose H^∞ norm is bounded by one) in the linear-fractional-transformation form

$$F(s) = K_{11}(s) + K_{12}(s)U(s)[I - K_{22}(s)U(s)]^{-1}K_{21}(s) \tag{7.72}$$

where $K_{ij}(s)$ are all computable from the given data. The theory for this result is too lengthy to develop here. Details may be found in the texts of Francis [65] or Vidyasagar [199]. A discussion of the two-Riccati-equation solution of the U-parameterization problem is given at the end of this section, and related software is discussed in Section 7.6. This free, bounded-real matrix, referred to as the **U-parameter**, can then be used to minimize a nominal LQG performance measure [55]. In this way, a given level of stability robustness can be guaranteed, while a nominal LQG performance measure is minimized. The LTR design approach can also be used to guarantee a level of robustness, but minimizing the nominal LQG performance measure then becomes a problem.

The minimization of a nominal LQG performance measure using the U-parameter requires some further parameterization of the $U(s)$ function since an analytic solution of the optimization problem in $U(s)$ is not available. One approach is to parameterize $U(s)$ in terms of some design parameter k that preserves the bounded-realness of $U(s)$. A simple example is $U(s) = S'\Lambda S$, where S is a fixed orthogonal matrix and Λ is a diagonal matrix with entries given by k_i, the components of the vector k. Then bounded realness of $U(s)$ is preserved as long as $|k_i| \leq 1$. Once $U(s)$ is parameterized in terms of a finite-dimensional vector k, the design procedure outlined in Section 6.3 for fixed-order compensators may be used to minimize a nominal LQG performance measure. The problem is then reduced to a nonlinear optimization problem in the vector k. We summarize next the steps required in a U-parameter design.

1. Use H^∞ theory to compute the matrices $K_{ij}(s)$ for the U-parameterization of all compensators $F(s)$ that satisfy the robust-stability condition in (7.53).
2. Select a parameterization of $U(s)$ that preserves bounded-realness. The simplest possible choice is a constant matrix such as $U(s) = S'\Lambda S$. More complex parameterizations can be built up in the product form

$$U(s) = \prod_{i=1}^{p} B_i(s) \tag{7.73}$$

   where $B_i(s)$ are simple parameterized bounded-real matrices. Since the product of bounded-real matrices is bounded real, bounded-realness is preserved. From the two-Riccati theory (see [58]), it is known that the order of the H^∞ compensator $F(s)$, for $U(s) = 0$, is the same as the order of the plant. As more complex matrices $U(s)$ are considered, the order of $F(s)$ is increased.

3. Use a nonlinear optimization approach, such as the gradient approach outlined in Section 6.3 for fixed-order compensators, to compute the U-parameterized compensator $F(s)$. Unfortunately, unlike the Q-parameter approach, the function to be minimized is not convex in general, and a gradient algorithm may converge on a poor local minimum. However, U-parameterization always guarantees a feasible (robustly stabilizing) starting point.

- **Mixed H^2/H^∞ design.** Mixed H^2/H^∞ design, sometimes referred to as **combined H^∞/LQG design**, is based on the fact that *if the following nonstandard Riccati equation*

$$0 = \tilde{A}\mathcal{Q} + \mathcal{Q}\tilde{A}' + \tilde{B}\tilde{B}' + \gamma^{-2}\mathcal{Q}\tilde{C}_2'\tilde{C}_2\mathcal{Q} \tag{7.74}$$

has a solution $\mathcal{Q} \geq 0$, then

$$\|\tilde{C}_2(sI - \tilde{A})^{-1}\tilde{B}\|_\infty \leq \gamma \tag{7.75}$$

and

$$\|\tilde{C}_1(sI - \tilde{A})^{-1}\tilde{B}\|_2 \leq (tr[\mathcal{Q}\tilde{C}_1'\tilde{C}_1])^{1/2} \tag{7.76}$$

Proof: A proof of (7.75) has already been given for $\gamma = 1$ in (7.55). To extend $\|C(sI-A)^{-1}B\|_\infty \leq 1$ to $\|C(sI-A)^{-1}B\| \leq \gamma$, one can simply replace C by C/γ, which explains (7.74). To demonstrate the inequality of (7.76), note that (7.74) implies

$$0 \geq \tilde{A}\mathcal{Q} + \mathcal{Q}\tilde{A}' + \tilde{B}\tilde{B}'$$

But we also have

$$\|\tilde{C}_1(sI - \tilde{A})^{-1}\tilde{B}\|_2 = (tr[Q\tilde{C}_1{}'\tilde{C}_1])^{1/2}$$

where Q satisfies the Lyapunov equation

$$0 = \tilde{A}Q + Q\tilde{A}' + \tilde{B}\tilde{B}'$$

From the discussion on guaranteed cost in Section 7.3, it then follows that $\mathcal{Q}$ is an upper bound on Q, i.e., $Q \leq \mathcal{Q}$. This then completes the demonstration of (7.76).

This result shows that the satisfaction of a Riccati equation can be used to guarantee an H^∞ norm bound and simultaneously an H^2 norm bound. The first bound can be used to ensure robust stability for unstructured perturbations, and the second bound can be used to minimize a guaranteed cost for a nominal LQG problem. Lagrange multiplier methods can be used to minimize $tr[Q\tilde{C}_1'\tilde{C}_1]$, subject to the constraint (7.74). The resulting optimization equations are a bit complex and are not repeated here. However, the equations may be found in [23], theorem 3.1, or [147], theorem 3.2, for fixed-order dynamical compensators.

Finally, we summarize some important formulas for the determination of a controller $F(s)$ that guarantees a closed-loop system with an H^∞ norm bounded by γ, when such a controller exists. In particular, it is shown in [58] that this problem can be reduced to the solution of **two decoupled Riccati equations**. Following the notation in [58], the system dynamics are written as

$$\begin{aligned} \dot{x} &= Ax + B_1 w + B_2 u \\ z &= C_1 x + D_{11} w + D_{12} u \\ y &= C_2 x + D_{21} w + D_{22} u \end{aligned} \tag{7.77}$$

where

- x is the system state
- w is the exogenous signal (reference signal, disturbance signal, etc.)
- u is the control input
- z is the controlled output
- y is the measured output

The objective is to find a compensator transfer function $F(s)$ that generates u from y such that

$$\| T_{zw}(s) \|_\infty < \gamma \tag{7.78}$$

where $T_{zw}(s)$ is the closed-loop transfer function from w to z. For convenience, the transfer function of the two-input–two-output system given in the form (7.77) is denoted by

$$G(s) = \left[\begin{array}{c|cc} A & B_1 & B_2 \\ \hline C_1 & D_{11} & D_{12} \\ C_2 & D_{21} & D_{22} \end{array} \right] \tag{7.79}$$

It was then shown in [58] that for the special case where

$$D_{12}'[C_1 \quad D_{12}] = [0 \quad I] \tag{7.80}$$

and

$$\begin{bmatrix} B_1 \\ D_{21} \end{bmatrix} D_{21}' = \begin{bmatrix} 0 \\ I \end{bmatrix} \tag{7.81}$$

a controller that results in a closed-loop transfer function $T_{zw}(s)$ satisfying the H^∞ norm given in (7.78) exists, if solutions to the two Riccati equations,

$$\begin{aligned} 0 &= A'X + XA + X(\gamma^{-2}B_1B_1' - B_2B_2')X + C_1'C_1 & (7.82) \\ 0 &= AY + YA' + Y(\gamma^{-2}C_1'C_1 - C_2'C_2)Y + B_1'B_1 & (7.83) \end{aligned}$$

exist, such that $X \geq 0$ and $Y \geq 0$ and $\rho(XY) < \gamma^2$, where $\rho(W)$ denotes the spectral radius of the matrix W. When these existence conditions are met, a particular controller is given by the transfer function

$$F(s) := \left[\begin{array}{c|c} \hat{A} & -ZL \\ \hline F & 0 \end{array}\right] \tag{7.84}$$

where

$$\begin{aligned} \hat{A} &= A + \gamma^{-2}B_1B_1'X + B_2F + ZLG_2 & (7.85) \\ F &= -B_2'X & (7.86) \\ L &= -YC_2' & (7.87) \\ Z &= (I - \gamma^{-2}YX)^{-1} & (7.88) \end{aligned}$$

Furthermore, a U-parameterization of the controller $F(s)$ as in (7.72) is given by

$$K(s) = \left[\begin{array}{c|cc} \hat{A} & -ZL & ZB_2 \\ \hline F & 0 & I \\ -C_2 & I & 0 \end{array}\right] \tag{7.89}$$

The $K_{ij}(s)$ required in (7.72) are formed by a partition of $K(s)$ given in (7.89), compatible with a $K_{11}(s)$ that has a size equal to $F(s)$.

In [77], formulas are presented for $K(s)$ without the special assumptions given in (7.80) and (7.81). Computer software for the computation of $K(s)$ is discussed in Section 7.6. It is interesting to note that the solution of the H^∞ control problem reduces to the solution of two decoupled Riccati equations just as in the LQG case.

7.5 Linear Matrix Inequality Methods

In this section we discuss briefly **linear matrix inequality (LMI)** theory and its application to multimodel guaranteed-cost problems. We follow the development in Boyd et al. [32] (see also the monograph by Boyd et al. [34]). By LMI we mean a matrix inequality of the form

$$F(x) = F_0 + \sum_{i=1}^{m} x_i F_i > 0 \tag{7.90}$$

where x is an unknown vector (with components x_i) and F_i are given, symmetric matrices. Note that the inequality (7.90) is a matrix inequality, meaning that the matrix $F(x)$ is positive definite. The set of vectors x for which $F(x) > 0$ is convex, so that any robust control problem with constraints of this type and a convex performance objective can be reduced to a convex-programming problem. As noted in the previous section, convex programming problems are particularly attractive since their optimal solutions are global and efficient algorithms exist for finding optimal solutions. Recently, some new algorithms referred to as **interior-point algorithms** have been developed to solve convex-programming problems (see, e.g., [150]). These algorithms appear to be especially efficient for solving problems with linear matrix inequalities.

A basic result in LMI theory is the following: *The matrix inequality*

$$\begin{bmatrix} Q & S \\ S' & R \end{bmatrix} > 0 \tag{7.91}$$

where Q and R are symmetric matrices, is true if and only if

$$\begin{aligned} R &> 0 \\ Q - SR^{-1}S' &> 0 \end{aligned} \tag{7.92}$$

Proof: Multiply the second row of the matrix in (7.91) on the left by R^{-1} and the second column on the right by R^{-1}. This results in the matrix

$$\begin{bmatrix} Q & SR^{-1} \\ R^{-1}S' & R^{-1} \end{bmatrix} \tag{7.93}$$

These elementary block-matrix, row- and column-operations preserve the inequality in (7.91). Now, to the first row of (7.93) add the second row multiplied on the left by $-S$. This yields the matrix

$$\begin{bmatrix} Q - SR^{-1}S' & 0 \\ R^{-1}S' & R^{-1} \end{bmatrix} \tag{7.94}$$

Finally, to the first column of (7.94) add the second column multiplied by $-S'$ on the right. This yields the matrix

$$\begin{bmatrix} Q - SR^{-1}S' & 0 \\ 0 & R^{-1} \end{bmatrix} \tag{7.95}$$

again with the inequality preserved. This block-diagonal matrix will be positive definite if and only if $R^{-1} > 0$ and $Q - SR^{-1}S' > 0$. Since $R^{-1} > 0$ is equivalent to $R > 0$, the proof is thus complete.

■

Note that inequality (7.91) is an LMI in S, while the matrix inequality $Q - SR^{-1}S' > 0$ is not. Indeed, this last inequality is **quadratic** in S. To see that (7.91) is equivalent to the LMI form given in (7.90), simply expand S into a sum of the form

$$S = \sum_i x_i S_i \tag{7.96}$$

where S_i are basis matrices for the matrix S.

The next result relates guaranteed-cost control to LMI. Consider the problem of finding a symmetric matrix $P > 0$ and a matrix K such that

$$(A - BK)'P + P(A - BK) + W < 0 \tag{7.97}$$

where matrices A, B, and $W > 0$ are given. Using the same arguments as those in the previous section on guaranteed-cost control, one can see that satisfying inequality (7.97) is the same as finding a feedback-control law $u = -Kx$ for the system $\dot{x} = Ax + Bu$ that guarantees the bound

$$\int_0^\infty x'(t)Wx(t)dt < x'(0)Px(0) \tag{7.98}$$

for all $x(0) \neq 0$ (see, for example, the bound given in (7.20) and its proof). Here we use strict inequalities to simplify the presentation.

Note also that if $P > 0$ and $W > 0$, then inequality (7.97) guarantees the asymptotic stability of the closed-loop system $\dot{x} = (A - BK)x$. This may be shown by using the Lyapunov function $V = x'Px$ and noting that (7.97) implies that $\dot{V} < 0$. Inequality (7.97) is not convex in P and K; however, with the substitutions (see [106])

$$\begin{aligned} P &= Y^{-1}; \\ K &= XY^{-1} \end{aligned} \tag{7.99}$$

inequality (7.97) may be written as

$$(A - BXY^{-1})'Y^{-1} + Y^{-1}(A - BXY^{-1}) + W < 0. \tag{7.100}$$

If (7.100) is multiplied on the right and left by Y, one obtains the inequality

$$YA' - X'B' + AY - BX + YWY < 0 \tag{7.101}$$

But from the basic LMI result, this can be written as

$$\begin{bmatrix} -YA' + X'B' - AY + BX & Y \\ Y & W^{-1} \end{bmatrix} > 0 \tag{7.102}$$

which is an LMI in the matrices X and Y.

We now apply the above result to the following multimodel, guaranteed-cost problem. Consider the problem of minimizing an upper bound on

$$E\{\int_0^\infty x'(t)Wx(t)dt\} \tag{7.103}$$

where the expectation is taken over random initial conditions $x(0)$ with $E\{x(0)x'(0)\} = I$, given a finite set of plant models

$$\dot{x} = A_i x + B_i u \tag{7.104}$$

and fixed state feedback $u = -Kx$. Problems of this type occur, e.g., when one wishes to design a fixed controller for a system that is linearized about a finite set of operating points. Using the results just derived, this problem is seen to be equivalent to the problem of minimizing

$$tr\{Y^{-1}\} \tag{7.105}$$

subject to the LMI constraints in X and Y

$$\begin{bmatrix} -YA_i' + X'B_i' - A_iY + B_iX & Y \\ Y & W^{-1} \end{bmatrix} > 0; \quad Y > 0 \quad (7.106)$$

It can be shown that $tr\{Y^{-1}\}$ is convex in Y, so that (7.105) and (7.106) constitute a convex-programming problem. The minimization of 7.105 can be reduced to the minimization of

$$tr\{Z\} \quad (7.107)$$

if the following LMI is added to the LMI's in 7.106

$$\begin{bmatrix} Z & I \\ I & Y \end{bmatrix} > 0 \quad (7.108)$$

Note that the system of LMI's given in (7.106) is indeed of the form $F(x)$ given in (7.90), if we think of $F(x)$ as a block-diagonal matrix made up of the blocks in (7.106). We will not pursue further the numerical solution of this convex-programming problem but will note that this is an otherwise difficult problem with no analytic solution. Indeed, the solution of the linear matrix inequalities in (7.106) guarantees the **simultaneous stabilization** of the plants given by (7.104) by a fixed controller $u = -Kx$. This simultaneous-stabilization problem itself has no general analytic solution.

7.6 MATLAB Software

The MATLAB functions **ltru** and **ltry** in the **Robust Control Toolbox** may be used to do LQG/LTR design. The former does loop recovery at the "input" end, with a fixed state-feedback matrix K_c; and the latter does loop recovery at the "output" end, with fixed filter gain K_f.

For the **ltru** function the input data matrices are

- $$sys := \left[\begin{array}{c|c} A & B \\ \hline C & D \end{array}\right]$$
- dim is the size of A
- K_c is the full-state feedback matrix

- Xi is the nominal plant-noise matrix Ξ
- Th is the measurement-noise matrix Θ
- r is a row vector of increasing fictitious-noise coefficients
- w is a row vector of frequencies for Bode plots of the singular values of the loop gain

The function **ltru** will produce Bode plots of the singular values of $H_{LQG}(j\omega)$ for each of the values in the vector r and return a controller $F(s)$, specified by the realization (af, bf, cf, df), for the last value of r specified in the vector r. The singular values of $H_{LQG}(j\omega)$ are stored in **svl**. One may also include in the input data the singular-value Bode plot of $H_{LQR}(j\omega)$, for comparison purposes, in the function **svk**. The command for **ltru** is written

$$[af, bf, cf, df, svl] = ltru(sys, dim, K_c, Xi, Th, r, w, svk)$$

where svk may be omitted. The dual of this function is given by **ltry** (see [46] for more details).

We consider the helicopter model to illustrate the LQG/LTR approach for designing for given gain margins.

Example 7.2 Consider again the helicopter of example 5.3. The LQG optimal cascade compensator for this problem was found in example 6.4 to be

$$\begin{aligned}
F(s) &= [af, bf, cf, df] \\
af &= \begin{bmatrix} -0.0175 & -0.1436 & 0.3852 & -26.3518 \\ 0.0084 & -17.6863 & -4.0536 & -13.9065 \\ 0.0010 & 0.0018 & -6.7274 & -33.2584 \\ 0 & 0.0031 & 1.0000 & -5.1191 \end{bmatrix} \\
bf &= \begin{bmatrix} 0.0158 & -0.2405 \\ 9.0660 & -0.1761 \\ 0.0091 & 0.2289 \\ -0.0031 & 0.0893 \end{bmatrix} \\
cf &= \begin{bmatrix} -0.0033 & 0.0472 & 14.6421 & 60.8894 \\ 0.0171 & -1.0515 & 0.2927 & 3.2469 \end{bmatrix}
\end{aligned}$$

and $df = 0_{2\times 2}$. Let us then compute the optimal loop-transfer function

$$H_{LQG}(s) = F(s)C(sI - A)^{-1}B$$

and by use of the MATLAB command **series**

$$[a, b, c, d] = series[A, B, C, D, af, bf, cf, df]$$

we obtain

$$a = \begin{bmatrix} a_{11} & a_{12} \\ a_{21} & a_{22} \end{bmatrix}$$

with

$$a_{11} = \begin{bmatrix} -0.02 & -0.005 & 2.4 & -32.00 \\ -0.14 & 0.44 & -1.3 & -30.00 \\ 0 & 0.018 & -1.6 & 1.2 \\ 0 & 0 & 1.0000 & 0 \end{bmatrix}$$

$$a_{12} = \begin{bmatrix} 0 & 0 & 0 & 0 \\ 0 & 0 & 0 & 0 \\ 0 & 0 & 0 & 0 \\ 0 & 0 & 0 & 0 \end{bmatrix}$$

$$a_{21} = \begin{bmatrix} 0 & 0.0158 & 0 & -13.7831 \\ 0 & 9.066 & 0 & -10.0906 \\ 0 & 0.0091 & 0 & 13.1179 \\ 0 & -0.0031 & 0 & 5.1191 \end{bmatrix}$$

$$a_{22} = \begin{bmatrix} -0.0175 & -0.1436 & 0.3852 & -26.3518 \\ 0.0084 & -17.6863 & -4.0536 & -13.9065 \\ 0.001 & 0.0018 & -6.7274 & -33.2584 \\ 0 & 0.0031 & 1.0000 & -5.1191 \end{bmatrix}$$

$$b = \begin{bmatrix} 0.14 & -0.12 \\ 0.36 & -8.6 \\ 0.35 & 0.009 \\ 0 & 0 \\ 0 & 0 \\ 0 & 0 \\ 0 & 0 \\ 0 & 0 \end{bmatrix}$$

$$c = \begin{bmatrix} 0 & 0 & 0 & 0 & -0.0033 & 0.0472 & 14.6421 & 60.889 \\ 0 & 0 & 0 & 0 & 0.0171 & -1.0515 & 0.2927 & 3.2469 \end{bmatrix}$$

and $d = 0_{2\times 2}$. This data is then used to plot the singular values of $I + H_{LQG}(s)$ versus frequency using the MATLAB command **sigma**.

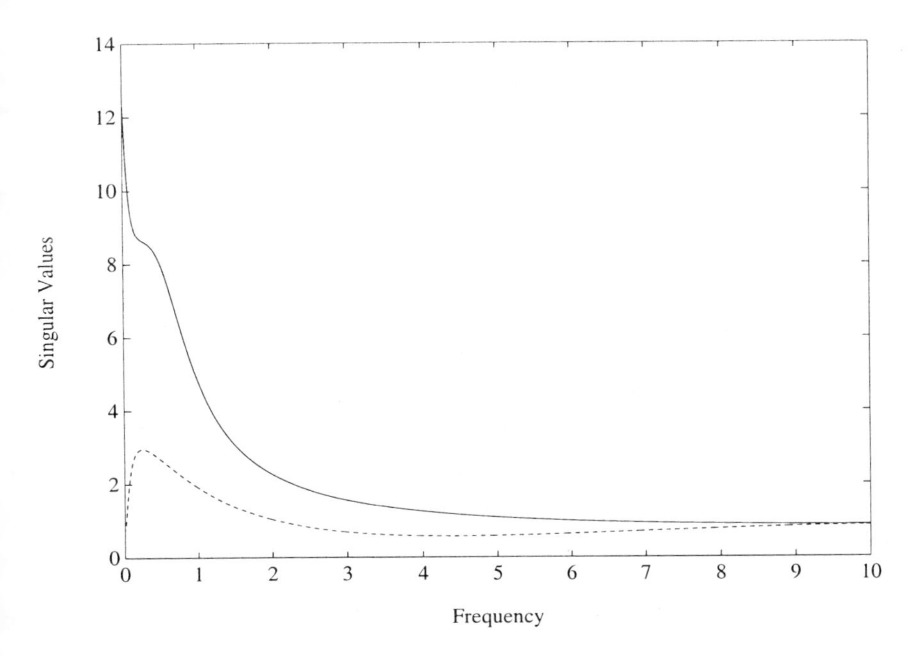

Figure 7.2: Singular Values of $I + H(jw)$ for the Helicopter Model with LQG Control

The plot of the singular values (two in this case) versus frequency is shown in Figure 7.2, from which we can determine the minimal singular value $\alpha = 0.57$. Using the results of problem 4.2, we obtain the gain margin in each input channel as

$$0.637 < l_i < 2.326$$

Consider now the problem of recovering the gain margins

$$0.606 < l_i < 2.857$$

which correspond to $\alpha = 0.65$. The MATLAB function **ltru** was used to find a value of fictitious noise corresponding to $\alpha = 0.65$, i.e., $r = 100$. The compensator corresponding to this value of r is given by

$$af = \begin{bmatrix} -0.0175 & 0.0084 & 0.0010 & 0 \\ -0.2479 & -22.5436 & 0.0006 & 0.0029 \\ 0.3852 & -4.0536 & -6.7274 & 1.0000 \\ -35.1906 & -14.3968 & -42.1594 & -6.6340 \end{bmatrix}$$

$$bf = \begin{bmatrix} -0.0033 & 0.0171 \\ 0.0472 & -1.0515 \\ 14.6421 & 0.2927 \\ 60.8894 & 3.2469 \end{bmatrix}$$

$$cf = \begin{bmatrix} 0.1202 & 13.9233 & 0.0103 & -0.0029 \\ -0.0863 & -0.1675 & 0.3843 & 0.1158 \end{bmatrix}$$

△

For guaranteed cost with linear bound, the same comments apply as for the solution of the stochastic algebraic equation (see Section 5.7). For the quadratic bound the function **are** can be used to solve the nonstandard Riccati equation. Finally, for the Chang-Peng bound, a sequence of standard Riccati equations must be solved. Each equation may be solved using the **lqr** function. An M-file is required to actually implement the algorithm in [41]. The following example illustrates the application of the quadratic-bound, guaranteed-cost design approach to the two-mass benchmark problem.

Example 7.3 Consider the two-mass spring problem presented in problem 2.13. Note that the design of robust controllers for this problem is discussed in detail in [1]. Assume that $m_1 = m_2 = 1$ and let $0.5 \leq k \leq 2.0$. Also assume that a guaranteed-cost controller is to be designed for this system, given

$$Q = \begin{bmatrix} 1 & 0 & 0 & 0 \end{bmatrix}' \begin{bmatrix} 1 & 0 & 0 & 0 \end{bmatrix}; \quad R = [1]$$

The state-space description for this system is given by

$$\dot{x} = \begin{bmatrix} 0 & 0 & 1 & 0 \\ 0 & 0 & 0 & 1 \\ -k & k & 0 & 0 \\ k & -k & 0 & 0 \end{bmatrix} x + \begin{bmatrix} 0 \\ 0 \\ 1 \\ 0 \end{bmatrix} u$$

With the transformation $k = 0.75w_1 + 1.25$ and the choices $D_1' = (0 \ \ 0 \ \ -1 \ \ 1)$ and $E_1 = (0.75 \ \ -0.75 \ \ 0 \ \ 0)$, the new A matrix for this system may be written as $A + w_1 D_1 E_1$, where

$$A = \begin{bmatrix} 0 & 0 & 1 & 0 \\ 0 & 0 & 0 & 1 \\ -1.25 & 1.25 & 0 & 0 \\ 1.25 & -1.25 & 0 & 0 \end{bmatrix}$$

and $-1 \leq w_1 \leq 1$. The quadratic-bound optimization equation may then be written

$$0 = A'P + PA + W - PVP$$

where

$$\begin{aligned} W &= Q + (1/\epsilon_1)^2 E_1' E_1 \\ V &= BR^{-1}B' - (\epsilon_1)^2 D_1 D_1' \end{aligned}$$

As noted in Section 3.6, the MATLAB function **are** can be used to obtain a solution of the nonstandard Riccati equation

$$0 = F'X + XF + H - XGX$$

where $G \geq 0$, but where H may be indefinite. In particular,

$$X = are(F, G, H)$$

returns the solution $X \geq 0$ when it exists. In our problem, the matrix G is indefinite. However, the guaranteed-cost equation can also be written as

$$0 = P^{-1}(-A') + (-A)P^{-1} + V - P^{-1}WP^{-1}$$

where $W \geq 0$ and V may be indefinite. If we let $P^{-1} = X$, the guaranteed cost solution may be obtained from

$$X = are(-A', W, V)$$

and $P = X^{-1}$. The state-feedback control law $u = -Kx(t)$ is given by

$$K = R^{-1}B'P$$

The parameter ϵ_1 may be considered a design parameter to help ensure the existence of a solution and a small-cost upper bound. We assume an initial state $x(0) = (1 \quad 0 \quad 0 \quad 0)'$. The guaranteed-cost bound is then given by

$$C_{GC} = \frac{1}{2} x'(0) P x(0)$$

Some computed values of C_{GC} for different values of ϵ_1 are shown in Table 7.1. The best choice of ϵ_1 from Table 7.1 is obviously $\epsilon_1 = 0.3$.

ϵ_1	C_{GC}
0.6	No Solution
0.4	20.58
0.3	7.93
0.2	9.11

Table 7.1: Guaranteed Cost for Some Values of ϵ_1

This choice results in the feedback matrix

$$K = R^{-1}B'P = \begin{bmatrix} 4.36 & -3.01 & 2.99 & 1.28 \end{bmatrix}$$

with guaranteed-cost bound equal to $C_{GC} = 7.93$. In general, the actual cost for a given guaranteed-cost controller will be less than the above bounds. Thus, for example, for the choice of K above, the actual cost at $w_1 = -1$ and $w_1 = 1$ are 2.28 and 2.9, respectively. This is considerably less than the guaranteed-cost bound of 7.93. It should be further noted that an LQR optimal controller designed for the nominal plant ($w_1 = 0$) may have robustness properties that are as good as or better than the guaranteed-cost design. This is the case for this particular example, where the nominal optimal gain

$$K = \begin{bmatrix} 0.75 & 0.25 & 1.22 & 0.67 \end{bmatrix}$$

yields cost values of 0.75 and 0.89 at the extreme values of $w_1 = -1$ and $w_1 = 1$.

△

While there are no specific MATLAB software packages that permit direct Q- or U-parameter design, several subroutines are available that expedite some of the computations required for these design techniques. The function **youla** may be used to compute the Q-parameterization of all nominally stabilizing compensators, and the function **hinf** may be used to compute the U-parameterization of all robustly stabilizing compensators. In particular, if the system dynamics are given in the form of (7.77), i.e., the input data are given by $A, B_1, B_2, C_1, C_2, D_{11}, D_{12}, D_{21}$, and D_{22}, the function **youla** in the MATLAB **Robust Control Toolbox** [46] generates the transfer functions $T_1(s), T_2(s)$, and $T_3(s)$ required in the Q-parameterization theory by outputting the matrices

$$at11, bt11, ct11, dt11,$$
$$at12, bt12, ct12, dt12,$$
$$dt21, bt21, ct21, dt21,$$

The required transfer functions are then given by

$$T_1(s) := \left[\begin{array}{c|c} at11 & bt11 \\ \hline ct11 & dt11 \end{array}\right]$$

$$T_2(s) := \left[\begin{array}{c|c} at12 & bt12 \\ \hline ct12 & dt12 \end{array}\right]$$

$$T_3(s) := \left[\begin{array}{c|c} at21 & bt21 \\ \hline ct21 & dt21 \end{array}\right]$$

These transfer functions are computed by the **youla** function, and they satisfy the following properties

$$T_2'(-s)T_2(s) = I; \quad T_3(s)T_3'(-s) = I$$

These special properties expedite the use of the affine term $T_1 + T_2QT_3$ for H^2 or H^∞ norm optimization. The function **augment** may then be used to put the equations of the sensitivity or complementary sensitivity matrices into the standard form given in equation (7.77).

The function **hinf** in the MATLAB **Robust Control Toolbox** [46] uses the two-Riccati-equation method to generate the matrices $K_{ij}(s)$ required in the U-parameter design. With the input data

$$A, B_1, B_2, C_1, C_2, D_{11}, D_{12}, D_{21}, D_{22}$$

the function **hinf** outputs the matrices

$$ak, bk1, bk2, ck1, ck2, dk11, dk12, dk22$$

The required $K_{ij}(s)$ are then given by the partition of the transfer function

$$K(s) := \left[\begin{array}{c|cc} ak & bk1 & bk2 \\ \hline ck1 & dk11 & dk12 \\ ck2 & dk21 & dk22 \end{array}\right]$$

As for the Q-parameter design approach, the function **augment** may be used to put equations into the standard form of (7.77). It should be noted, however, that the two-Riccati-equation theory requires that the matrix D_{12} be of full rank. For the H^{∞}-control problem represented by (7.53), which corresponds to a case where the weighting matrices W_1, W_2, and W_3 in the **augment** function are given by $W_1 = W_2 = 0$, $W_3 = \alpha I$, the full-rank condition on D_{12} is not met unless the plant is exactly proper. One way to get around this difficulty is to let $W_2 = \epsilon I$ and make ϵ very small.

The function **lftf** may be used to compute linear-fractional transformations, such as (7.72), for a given parameter matrix $U(s)$. The functions **are** and **lyap** may be used to solve Riccati and Lyapunov equations. Finally, the **Optimization Toolbox** [79] contains a number of nonlinear programming algorithms that may be used in the final optimization steps of Q- and U-parameter design.

Finally the function **mincx** in the **LMI Toolbox** [72] can be used to minimize a linear trace function of the type given in equation 7.107, subject to LMI contraints such as in 7.106 and 7.108.

7.7 Notes and References

An excellent survey of the LQG/LTR design approach, with applications, may be found in Stein and Athans [190]. See also the text of Saberi, Chen, and Sannuti [175] . In Chen, Saberi, and Sannuti [43], exact and asymptotic loop-transfer recovery is explored in some detail. A nonobserver-based compensator (Kalman-Bucy) is proposed to obtain loop recovery with lower levels of gain. Applications of the LQG/LTR approach to aerospace problems may be found in Athans et al. [13] and Ridgely et al. [173]. For additional references on loop recovery, see [27], [42], [133], [145], and [183].

The problem of finding constant matrices P such that $V = x'Px$ is a Lyapunov function for an uncertain system is referred to as the **quadratic-stabilization problem**. More details on the quadratic stabilization problem may be found in [15], [22], [163], [165], [166], and [215]. In particular, in [22] the quadratic-stabilization problem with static, output-feedback and fixed-order compensators is also considered.

In this chapter we considered only guaranteed cost with state feedback. The problem of guaranteed cost with observer feedback was first considered in Goldstein [78]. The problem of guaranteed-cost control with fixed-order compensators is studied by Bernstein and Haddad in [24] and [25]. As noted in Chapter 6, however, very little theory is available for the fixed-order problem since even the problem of nominal fixed-order stabilization is an open problem.

An introduction to H^∞ control theory, using interpolation theory, may be found in the texts of Dorato, Fortuna, and Muscato [51] and Doyle, Francis, and Tannenbaum [57]. More advanced material on the subject, using operator theory, may be found in Francis [65]. In 1989, Doyle et al. [58] showed that the H^∞ control problem could be reduced, as in the LQG case, to the solution of two decoupled Riccati equations. The Riccati equations were, however, of the nonstandard type. This was a significant development in H^∞ control theory, since the solution was reduced to well-known matrix computations, and a bound on the order of the compensator was established. In the case of multiplicative robustness design (see (7.48)), the order was simply the order of the plant plus the order of the bounding function $r(s)$.

In 1976, Youla, Bongiorno, and Jabr [213] presented a parameterization for multivariable systems of all compensators that yield a nominally stable closed-loop system. This parameterization is now commonly referred to as Q-parameterization or Youla parameterization. This same parameterization was independently developed by Kučera [110] for discrete-time systems. In 1988, Boyd et al. [31] proposed the approach to the design of linear feedback systems based on the convex programming and Q-parameterization we have outlined in Section 7.4. More details on this design approach may be found in the text by Boyd and Barratt [33].

U-parameter design theory was first presented for single-input–single-output systems in Dorato and Li [53]. U-parameter theory for the multivariable LQG problem is developed in Dorato and Yen [55]. More details and a numerical example may be found in this reference.

The mixed H^2/H^∞ approach outlined in Section 7.4 was developed by Bernstein and Haddad in 1989 [23] for the general case of fixed-order compensators. In Mustafa [147], the role of the Lagrangian-multiplier matrix $\mathcal{P}$ for problems of this type is explored in detail, and in Mustafa [148], the relationship between maximum-entropy/H^∞ control (Mustafa and Glover [149]) and mixed H^2/H^∞ control is established. In Khargonekar and Rotea [106] it is shown that for the mixed H^2/H^∞ problem, state feedback is sufficient for an optimal solution; i.e., if the state of the system is available the compensator need not be dynamic. It is also shown in [106] that the mixed H^2/H^∞ can be reduced to a convex optimization problem. The more complex problem of H^2 guaranteed-cost design, subject to a robust H^∞ bound constraint is considered in Madiwale, Haddad, and Bernstein [127]. This problem is reduced in [127] to a coupled system of three modified Riccati equations and a modified Lyapunov equation.

It should be noted that minimizing an upper bound on the H^2 norm, as is done in the mixed H^2/H^∞ approach, is not the same as actually minimizing the H^2 norm, as is done in the Q- and U-parameter approaches. Indeed, it is possible that a minimal upper-bound solution may be a poor solution to minimal norm solution. In Rotea and Khargonekar [174], sufficient conditions are given for the more difficult problem of minimizing an H^2 norm subject to an H^∞ constraint. It is interesting to note (see [174]), that for this case the compensator must be dynamic even if the state of the system is available for feedback. In Saberi et al. [176], the singular case is considered.

For certain mechanical problems, with co-located sensors and actuators, the plant is "passive" for all possible parameter variations. Passive means that the plant transfer function is a **positive real matrix**. A matrix $Z(s)$ is positive real if: (1) $Z(s)$ is real for s real, (2) $Z(s)$ is analytic for $\mathrm{Re}s > 0$, and (3) $Z(s)+Z^*(s)$ is positive semidefinite in $\mathrm{Re}s \geq 0$, where $Z^*(s)$ denotes the complex-conjugate transpose of $Z(s)$. For this class of problems, which includes some interesting large-space-structure problems, any passive compensator will guarantee robust stability of the closed-loop system. For LQR design for this class of robust control problems see Joshi and Maghami [98] and McLaren and Slater [136].

In our discussion of robust design for linear-quadratic problems, we have always assumed that the Q and R matrices were assigned and non-negotiable. Of course, this is not always the case. In Trofino-Neto, Dion, and Dugard [194], techniques are presented for the adjustment of the Q and R matrices to ensure that the solution of a nominal LQ problem guarantees a robustly stable closed-loop system in the presence of bounded-parameter variations. In Vinkler and Wood [201], the Q and R matrices are tuned in the guaranteed-cost approach to reduce feedback gains and improve the transient response.

It should be noted that to simplify the presentation we have assumed that perturbations occurred only in the A matrix in our state-space plant models. If perturbations occur in the input matrix B, the methods presented are still applicable, but the resulting design equations are considerably complicated. See Petersen [165] for a discussion of robust stabilization in the case of input-matrix perturbations.

Other references that discuss the robustness problem in linear-quadratic control are [83], [92], [117], [152], [141], [159], [185], and [191].

7.8 Problems

Problem 7.1 For the LQG example discussed in Section 7.1, do a Nyquist plot of the nominal ($m = 1$) loop gain ($H_{LQG}(j\omega)$), where $\alpha = \beta = 10$. Use the MATLAB function **nyquist** if necessary. What is the phase margin in this case?

Problem 7.2 Use the LQG/LTR design approach to find a compensator $F(s)$ that recovers a phase margin of 30 degrees for problem 7.1. Hint: Recall that the guaranteed phase margin is related to the minimal value of $\underline{\sigma}(I + H_{LQG}(j\omega))$ (see problem 4.2). To solve this problem, you will need to use some LQG/LTR software and obtain plots of $\underline{\sigma}(I + H_{LQG}(j\omega))$.

Problem 7.3 Consider the plant

$$\dot{x} = \begin{bmatrix} 0 & 1 \\ 0 & w_1 \end{bmatrix} x + \begin{bmatrix} 0 \\ 1 \end{bmatrix} u; \quad |w_1| \leq 1$$

and an LQR design problem with $Q = I_{2\times 2}$ and $R = [1]$. Set up the guaranteed-cost optimization equations for linear and quadratic bounds. Obtain, if they exist, solutions to the optimization equations for $\gamma_1 = 1$, $\epsilon_1 = 1$, $E_1 = [0 \ \ 1]$ and $D_1 = E_1'$.

Problem 7.4 Consider a guaranteed-cost design problem with uncertainty in only the input matrix B, i.e., $\Delta A = 0$ but $\Delta B = w_1 B_1$. Derive the optimization equation for this case using a linear bound. Hint: Show that the required bounding function in this case is given by

$$\mathcal{U}(P) = \frac{1}{\gamma_1} P + \gamma_1 K' B_1' P B_1 K$$

and note that $\mathcal{U}(P)$ is now also a function of the design parameter K. Use this result to design a guaranteed-cost system for the LQR state-feedback problem discussed in Section 7.1, where

$$A = \begin{bmatrix} -1 & 0 \\ 0 & -2 \end{bmatrix}$$

and $Q = [1 \quad -1]'[1 \quad -1], B = [1 + w_1 \quad 1]', R = [0.01]$. Explore different choices of γ_1, e.g. $\gamma_1 = 1, 10$.

Problem 7.5 Consider the equations

$$\begin{aligned} \dot{x} &= Ax + B_1 u + B_2 w \\ z_1 &= x \\ z_2 &= u \end{aligned}$$

Assume state feedback of the form $u = -Kx$. Derive the optimization equation required for the minimization on an upper bound on the H^2 norm of the transfer matrix between exogenous input w and the output z_1 subject to the H^∞ norm of the transfer matrix from w to z_2 being bounded by γ. Hint: Minimize the Lagrangian

$$\mathcal{L} = tr[\mathcal{Q}\tilde{C}_1'\tilde{C}_1 + (\tilde{A}\mathcal{Q} + \mathcal{Q}\tilde{A}' + \tilde{B}\tilde{B}' + \gamma^{-2}\mathcal{Q}\tilde{C}_2'\tilde{C}_2\mathcal{Q})\mathcal{P}]$$

where $\tilde{A} = A - B_1 K, \tilde{B} = B_2, \tilde{C}_1 = I$, and $\tilde{C}_2 = K$. Solve the optimization equation for the scalar problem $A = [1], B_1 = B_2 = [1]$. For what values of γ does a solution exist?

Problem 7.6 Consider the system given in example 1 in [41]:

$$\dot{x} = (A + w_1 D_1 E_1)x + Bu; \quad | w_1 | \leq 1$$

where

$$A = \begin{bmatrix} 0 & 1 & 0 \\ 0 & 0 & 1 \\ 0 & 0 & 0 \end{bmatrix}; \quad D_1 = \begin{bmatrix} 0 \\ 0 \\ 1 \end{bmatrix}; \quad E_1 = 1.6 \begin{bmatrix} 0 & 0 & 1 \end{bmatrix}$$

$$B = \begin{bmatrix} 0 \\ 0 \\ 1 \end{bmatrix}$$

with performance-measure matrices

$$Q = \begin{bmatrix} 1 & 0 & 0 \\ 0 & 0 & 0 \\ 0 & 0 & 0 \end{bmatrix}; \quad R = [1]$$

1. Use the quadratic bound to compute cost bounds when they exist, for $\epsilon_1 = 1.0,\ 0.9$, and 0.8. Find the feedback-gain matrix K corresponding to the best choice of ϵ_1.

2. Compute the optimal gain for the nominal system ($w_1 = 0$). Verify that the guaranteed-cost gain yields a robustly stable feedback system for all $\mid w_1 \mid\leq 1$. Is the nominally optimal gain robustly stable?

Problem 7.7 Consider the active suspension discussed in problem 6.6.

1. Design an LQG/LTR controller for the data in Table 6.1 when a fictitious plant noise is added such that $r = 100$. Plot the singular values of $I + H_{LQG}(s)$ for this controller and compute the corresponding increasing and decreasing guaranteed gain margins (see problem 4.2).

2. Repeat (1) for an LQG controller, i.e., an LQG/LTR controller with $r = 0$.

3. Assume that the vehicle mass deviates from its nominal value of $M = 250$ kg. For what range of values of M does the LQG/LTR closed-loop system designed in (1) remain stable? For what range of values of M does the LQG system designed in (2) remain stable? Note that the stability of the closed-loop system is determined by the eigenvalues of the matrix F in (6.40). This problem is discussed in some detail in [170].

Problem 7.8 Consider the helicopter model of example 5.3. Assume the same data as in that example except that the $(2, 2)$ term in the A matrix is now assumed to be uncertain but known to lie in the interval $[0.24, 0.64]$.

1. With Q and R the same as in (1) of example 5.3, explore a guaranteed-cost design for this problem, using a quadratic upper bound. Hint: The unknown coefficient must be mapped into a variable w that varies in the interval $[-1, 1]$, and matrices D_1 and E_1 must be selected. Finally, (7.45) must be solved. Select a mapping such that the nominal value of 0.44 corresponds to $w = 0$. Select D_1 to be an appropriate column matrix and E_1 to be an appropriate row matrix. Finally,

explore the effect of different values of ϵ_1 on the solution of (7.45).

2. Compare the solution of the guaranteed-cost problem in (1) above with an LQR design based on the nominal value. In particular, compare the actual performance-measure value in both designs for the $(2,2)$ entry values of 0.24, 0.44, and 0.64.

Problem 7.9 Show that:

1.

$$\begin{bmatrix} Q & S & T \\ S' & W & 0 \\ T' & 0 & V \end{bmatrix} > 0$$

if and only if

$$W > 0; \quad V > 0; \quad Q - SW^{-1}S' - TV^{-1}T' \ > \ 0$$

2. Use the result in (1) to derive convex-optimization equations for the multimodel guaranteed-cost problem, as described in Section 7.5, when the performance measure is

$$E\{\int_0^\infty (x'Wx + u'Vu)dt\}$$

instead of (7.103).

Chapter 8

Digital Control

Since most controllers are implemented digitally, usually with microprocessors, we outline briefly in this chapter how the linear-quadratic theory developed here for analog control may be extended to digital systems. We limit our discussion here to the control of analog plants.

Of course, one immediate possibility is simply to do an analog design and then to replace the analog compensator with a digital filter, with appropriate analog-to-digital (A/D) and digital-to-analog (D/A) converters to interface the analog plant with the digital filter. The problem then is that of approximating an analog filter with a digital filter. This is a digital-filtering problem, and a great deal of literature is available on the subject, see, e.g., the classic text of Oppenheim and Schafer [153]. In Keller and Anderson [105], the effects of digital controller approximation on the closed-loop system behavior is explored in some detail. We will not explore this approach any further here.

Another approach is to replace the analog plant by a suitable discrete-time system, which characterizes the evolution of the plant at the sampling instants. Digital control invariably involves sampling of signals, and indeed an alternative term used for **digital control** is **sampled-data control**. One problem with this approach is that it is difficult to control the behavior of the analog plant between the sampling times. We present an approach here that for simple sample-and-hold on the plant input reduces the control problem to a discrete-time problem while including in the performance index the behavior of the plant output at all values of time. This approach leads to a discrete-time, linear-quadratic optimization prob-

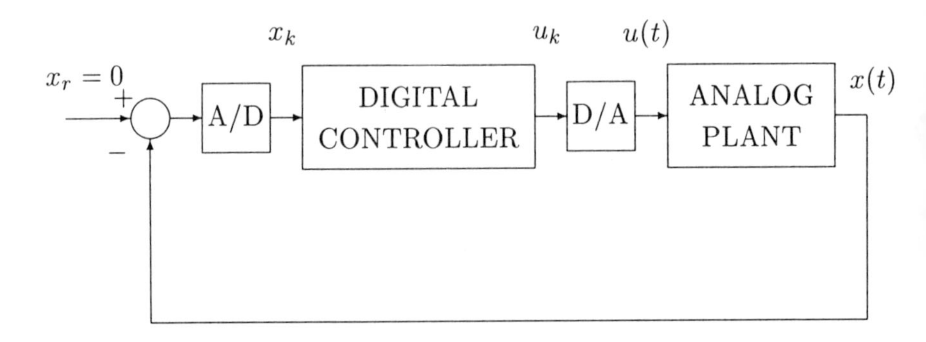

Figure 8.1: Digital-Control Structure

lem. Discrete-time dynamic programming is then used to develop an appropriate optimization equation. The extension of the theory developed in the previous chapters for analog systems to digital systems is, in most cases, fairly direct. Thus, we do not repeat in this chapter all the detailed results presented for analog systems, e.g., tracking and disturbance rejection, stochastic control, robustness design, etc. Rather, we present in some detail only the discrete-time theory for state-feedback LQR control. It should be noted that not all analog results generalize easily to digital results. Indeed, some digital results are unique to digital systems, e.g., dead-beat response, and have no analog counterpart. Some of the important differences between analog and digital theory are discussed in Dorato [50].

8.1 Digital-Control Issues

A detailed study of all the issues associated with digital control is beyond the scope of this text. The reader is referred to specialized books on the subject, such as the monograph of Moroney [146]. It is also assumed that the reader is familiar with the basic theory of discrete-time systems, such as found for example in Franklin, Powell, and Workman [70] or Lewis [121]. However, we review briefly some of the more important issues in this section. The basic digital-control structure we will consider is shown in Figure 8.1.

- **Sampling-time selection**. This is a very important issue in digital control and involves many tradeoffs. Of course, systems with "fast" time responses require small sampling intervals. On the other hand, small sampling times require more expensive A/D, D/A, and digital computers. In our presentation we assume the sampling time, denoted T, has been fixed. See [48] for a discussion of multirate sampling for LQG problems.

- **Amplitude quantization**. Since digital computers use only finite arithmetic, all signals must be quantized in amplitude. Here too we have tradeoffs to make. We will assume that the quantization level is fine enough that the signals may be assumed to be continuous in amplitude. For highly accurate control, this assumption may not hold, and a nonlinear design may be required so that finite amplitude quantization can be dealt with. The design of such systems is beyond the scope of this text, however.

- **Computation time**. After the state of the system is measured, some computation time, denoted δ, is always required to compute the control law over the next sampling interval. Even with a "fast" computer this interval may be a significant fraction of the required sampling interval for a given problem. We will assume, however, that the computation time is negligible. A brief discussion of how to allow for computation may be found in sect. 11.4 of [4].

- **Anti-aliasing analog prefilter**. Generally, an analog filter is required ahead of the A/D converter to prevent "aliasing" effects that occur when one samples a signal that is not sufficiently band limited. See sect. 11.3 of [4] for a discussion of this phenomenon and its correction by suitable prefiltering. We will assume that any such filter has already been selected and its dynamics incorporated into the plant.

8.2 Discrete-Time LQR Problem

In this section we show how an analog LQR problem can be reduced to a discrete-time LQR problem when the control input is held constant over each sampling interval (zero-order hold). We assume that the analog plant is characterized by the continuous-time equation

$$\dot{x} = \tilde{A}x + \tilde{B}u \tag{8.1}$$

that the control input is constant over any sampling interval, i.e., $u(t) = u_k$ for $kT \leq t < (k+1)T$; and that the state of the system is available at the sampling times kT, where k is an integer-valued variable, and T is the sampling interval. For simplicity we write $x_k = x(kT)$ for the state and other signals sampled at time kT. We further assume, as in [52], that the object is to find a digital compensator that generates the control values u_k such that the **continuous-time**, integral-quadratic performance measure

$$V = \int_t^{t_f} (x'\tilde{Q}x + u'\tilde{R}u)d\tau + x'(t_f)Sx(t_f) \tag{8.2}$$

is minimized. Here we use t_f to denote the (fixed) final time and S to denote the terminal-cost matrix. The matrices $\tilde{A}, \tilde{B}, \tilde{Q}$, and $\tilde{R}$ may be time-varying; however, for simplicity of notation we will omit the variable t in these matrices. If we let $\Phi(t, \tau)$ denote the transition matrix for the system $\dot{x} = \tilde{A}x$, then for given x_k and u_k we have

$$x(t) = \Phi(t, kT)x_k + \int_{kT}^{t} \Phi(t, \tau)\tilde{B}u_k d\tau, \quad kT \leq t < (k+1)T \tag{8.3}$$

and

$$x_{k+1} = A_k x_k + B_k u_k \tag{8.4}$$

where

$$A_k = \Phi((k+1)T, kT), \quad B_k = \int_{kT}^{(k+1)T} \Phi((k+1), \tau)\tilde{B}d\tau \tag{8.5}$$

The difference equation (8.4) characterizes the evolution of the state at the discrete sampling times $t = kT$. Consider now the performance measure (8.2) over a sampling interval. With (8.3) one has

$$\int_{kT}^{(k+1)T} (x'\tilde{Q}x + u'\tilde{R}u)d\tau = x_k'Q_k x_k + 2u_k'M_k x_k + u_k'R_k u_k \tag{8.6}$$

where

$$Q_k = \int_{kT}^{(k+1)T} \Phi'(\tau, kT)\tilde{Q}\Phi(\tau, kT)d\tau \tag{8.7}$$

$$M_k = \int_{kT}^{(k+1)T} H_k'(\tau)\tilde{Q}\Phi(\tau, kT)d\tau \tag{8.8}$$

$$R_k = \int_{kT}^{(k+1)T} [\tilde{R} + H_k'(\tau)\tilde{Q}H_k(\tau)]d\tau \tag{8.9}$$

where

$$H_k(\tau) = \int_{kT}^{T} \Phi(\tau, \alpha)\tilde{B}d\alpha \tag{8.10}$$

From (8.6) it then follows that the integral-quadratic performance measure (8.2) can be written as the sum

$$V(x_i, i) = \sum_{k=i}^{N-1} l(x_k, u_k, k) \;\; + x'_N S x_N \tag{8.11}$$

where $t_f = NT$ and

$$l(x, u, k) = x' Q_k x + 2u' M_k x + u' R_k u \tag{8.12}$$

Note that no approximations were made in reducing the integral in (8.2) to the sum in (8.11), only the assumption that the control input $u(t)$ is piecewise constant. For notational simplicity we will drop the subscript k in the matrices A_k, B_k, Q_k, etc., even if the matrices may depend on k. The results that follow hold for either constant matrices or time-varying matrices, i.e., matrices that depend on k.

8.3 Discrete-Time LQR Optimization

We now use the optimality principle to derive an optimization equation for the discrete-time problem corresponding to the performance measure (8.11) and the dynamics (8.4). Let $V^*(x_i, i)$ denote the minimal value of performance starting at time $t = iT$ and state $x(iT) = x_i$. Then the optimality principle states that any input that is optimal over the interval (i, N) must necessarily be optimal over the interval $(i+1, N)$ so that the following recursive relation must hold true

$$V^*(x_i, i) = \min_{u_i}[l(x_i, u_i, i) + V^*(x_{i+1}, i+1)]; \quad i = N-1, N-2, ... \tag{8.13}$$

where $x_{i+1} = Ax_i + Bu_i$. Equation (8.13) is the optimization equation we are seeking. It is the discrete-time analog of the Hamilton-Jacobi equation derived in Section 2.2. In the discrete-time case the optimization equation can be solved by starting at the end time N

and optimizing backward in time step by step. Thus, e.g., at time $i = N$ if the state x_N is known we have

$$V^*(x_N, N) = x'_N S x_N \tag{8.14}$$

Now we move back one step in time and assume x_{N-1} is known. Then

$$V^*(x_{N-1}, N-1) = \min_{u_{N-1}} [l(x_{N-1}, u_{N-1}, N-1) + V^*(x_N, N)] \tag{8.15}$$

where $V^*(x_N, N) = x'_N S x_N$ and $x_N = Ax_{N-1} + Bu_{N-1}$. The value of u_{N-1} that minimizes the right-hand side of (8.15) may be computed by taking the gradient with respect to u_{N-1} and setting it equal to zero. This results in the following optimal value for u_{N-1}

$$u^*_{N-1} = -(R + B'SB)^{-1}(M + B'SA)x_{N-1} \tag{8.16}$$

With the value of u_{N-1} fixed we now move back to $i = N - 2$ and optimize with respect to u_{N-2}, etc. It is not difficult to see that at each step $V^*(x_i, i)$ will be quadratic in x_i. Indeed, if we let $V^*(x_i, i) = x'_i P_i x_i$ we can use (8.13) to obtain a difference equation for the matrix P_i, which we will refer to as the discrete-time Riccati equation. In particular, with this substitution we have

$$\begin{aligned} x'_i P_i x_i &= \min_{u_i} [x'_i Q x_i + 2u'_i M x_i + u'_i R u_i \\ &\quad + (Ax_i + Bu_i)' P_{i+1} (Ax_i + Bu_i)] \end{aligned} \tag{8.17}$$

Performing the indicated minimization with respect to u_i, by setting the gradient equal to zero in the usual way, we obtain $u^*_i = -K_i x_i$, where

$$K_i = (R + B'P_{i+1}B)^{-1}(B'P_{i+1}A + M) \tag{8.18}$$

If this value of u_i is substituted back into (8.17) one obtains after some matrix manipulations the **discrete-time Riccati equation**

$$\begin{aligned} P_i &= (Q + A'P_{i+1}A) \\ &- (M + B'P_{i+1}A)'(R + B'P_{i+1}B)^{-1}(M + B'P_{i+1}A) \end{aligned} \tag{8.19}$$

which may be solved recursively, backward in time, starting with

$$P_N = S \tag{8.20}$$

The optimal control is then given in state-feedback form $u^*_i = -K_i x_i$, where K_i is given by (8.18) and where P_i is a solution of the discrete-time Riccati equation, (8.19).

8.4 Steady-State Control

In this section we study the case where all "discrete-time" matrices A, B, Q, M, and R are time-invariant, i.e., independent of k, and N approaches infinity. Note that if the original "analog" matrices $\tilde{A}, \tilde{B}, \tilde{Q}$ etc., are time-invariant, i.e., independent of t, then the discrete-time matrices will be independent of k. Thus, for example, if $\tilde{A}$ and $\tilde{B}$ are constant, then $\Phi(t,\tau) = e^{\tilde{A}(t-\tau)}$ and A is given by

$$\begin{aligned} A &= \Phi((k+1)T, kT) \\ &= e^{\tilde{A}T} \end{aligned} \tag{8.21}$$

while

$$B = \int_0^T e^{\tilde{A}\alpha}\tilde{B}d\alpha \tag{8.22}$$

which are clearly independent of k. Note incidently that the discrete-time matrix A is always nonsingular, since a matrix exponential is always nonsingular. One can also show that all the other discrete-time matrices are independent of k. In particular, note that if we let

$$H(t) = \int_0^t e^{\tilde{A}\beta}\tilde{B}d\beta \tag{8.23}$$

then

$$\begin{aligned} H_k(\tau) &= \int_{kT}^{\tau} e^{\tilde{A}(\tau-\alpha)}\tilde{B}d\alpha \\ &= \int_0^{\tau-kT} e^{\tilde{A}\beta}\tilde{B}d\beta \\ &= H(\tau - kT) \end{aligned} \tag{8.24}$$

This indicates that $H_k(\tau)$ is a function only of the difference $(\tau - kT)$. This is also true of $\Phi(\tau, kT)$ when $\tilde{A}$ is constant, so that Q, M, and R can be computed from

$$\begin{aligned} Q &= \int_0^T e^{\tilde{A}'\beta}\tilde{Q}e^{\tilde{A}\beta}d\beta \\ M &= \int_0^T H'(\beta)\tilde{Q}e^{\tilde{A}\beta}d\beta \\ R &= \int_0^T (\tilde{R} + H'(\beta)\tilde{Q}H(\beta))d\beta \end{aligned} \tag{8.25}$$

The above integrals result from (8.7 through 8.9) and the change of variables $\beta = \tau - kT$ (see sect. 9.44 of [70], for a discussion of the numerical computation of the above integrals).

It is also possible to minimize a performance measure that involves the state and control input only at the sampled times, i.e.,

$$V_s = \sum_{k=0}^{N-1} (x_k' \tilde{Q} x_k + u_k' \tilde{R} u_k) \tag{8.26}$$

For this case we have $Q = \tilde{Q}$, $R = \tilde{R}$ and $M = 0$. The formulas (8.27) and (8.28) may be used to compute the optimal feedback-control law $u_k^* = -Kx_k$. This approach does not take into account the behavior of the system between sampling times. However, for small-enough sampling intervals T, the sampled performance measure V_s may be an acceptable measure for the performance of the analog plant.

If the solution of the discrete-time Riccati equation has a limit as N approaches infinity, i.e.,

$$\lim_{N\to\infty} P_i = \bar{P}$$

then $\bar{P}$ must satisfy the **discrete-time algebraic Riccati equation**

$$\begin{aligned}\bar{P} &= (Q + A'\bar{P}A) \\ &\quad -(M + B'\bar{P}A)'(R + B'\bar{P}B)^{-1}(B'\bar{P}A + M)\end{aligned} \tag{8.27}$$

and the feedback-control law becomes independent of i, i.e., $u_i^* = -Kx_i$, where K is given by

$$K = (R + B'\bar{P}B)^{-1}(B'\bar{P}A + M) \tag{8.28}$$

We explore next the conditions required for the steady-state solution to exist. The theory for the steady-state, discrete-time regulator problem parallels very closely the continuous-time theory, developed in Section 2.5, so we only briefly outline the main results.

Existence and Stability of the Steady-State, Discrete-Time LQR Solution. *Given the discrete-time LQR problem defined by the performance measure (8.11) and dynamics (8.4) with $S = 0$, $R > 0$, $Q - M'R^{-1}M \geq 0$, and $Q = D'D$, where the pair (A, D) is observable and the pair (A, B) is controllable, it follows that a solution to* *the steady-state LQR problem exists, in particular, that there exists a unique positive-definite solution to discrete-time ARE (8.27), $\bar{P}$, and that the optimal closed-loop system, i.e., $x_{k+1} = (A - BK)x_k$, where K is given by (8.28), is* **asymptotically stable**. ■

Because of the similarity to the continuous-time case we will not prove the above result here; however, we will comment on some of the particulars of some of the discrete-time issues.

- The condition $Q - M'R^{-1}M \geq 0$ is required for the discrete-time problem to guarantee that the function $l(x_i, u_i, i)$ is nonnegative (see problem 2.8).

- The conditions of observability and controllability may be relaxed, as in the continuous-time case, to detectability and stabilizability.

- With A nonsingular, the conditions for observability and controllability in the discrete-time case are identical to those in the continuous-time case; e.g., the discrete-time pair (A, B) is controllable if and only if

$$\text{rank}[B|AB|\ldots|A^{n-1}B] = n$$

  However, for pure discrete-time problems where A may be singular, conditions for controllability and observability are more complicated (see, e.g., [52]).

- *The discrete-time system $x_{k+1} = Fx_k$ is asymptotically stable if and only if all the eigenvalues of the matrix F have magnitudes strictly less than one.* If the matrix F is nilpotent, i.e., $F^\nu = 0$ for some integer ν, then the system $x_{k+1} = Fx_k$ is said to be a **dead-beat system,** since $x_k = F^k x_0 = 0$ for all $k \geq \nu$. It is possible that an optimal steady-state, discrete-time solution will yield a dead-beat, closed-loop response. This can never happen in the continuous-time case, since a matrix exponential can never be nilpotent.

- In the discrete-time case the condition $R > 0$ can be relaxed to $R \geq 0$, as long as $(R + B'\bar{P}B)$ is nonsingular.

- The discrete-time ARE can be numerically solved in ways that are analogous to the continuous-time case, i.e., approximation-in-policy-space, eigenvalue-eigenvector methods, etc. (see, [52] for a summary of numerical methods for the discrete-time ARE).
- It is possible that with sampling, the discrete-time system may not be controllable even when the analog plant is controllable (for a discussion of the possible loss of controllability see [52]).

Example 8.1 Consider the problem of digital control for the analog system $\dot{x} = u$ with performance measure

$$V = \int_0^\infty (x^2 + u^2)dt$$

for the sampling interval $T = 1$.

Solution: For this example we have

$$\tilde{A} = 0;\ \tilde{B} = 1;\ \tilde{Q} = 1;\ \tilde{R} = 1$$

The transition matrix corresponding to $\dot{x} = 0$ is given by $\Phi(t, \tau) = 1$. Using this function in the expressions for the discrete-time matrices A, B, Q, etc., given by (8.5) and (8.7) through (8.10), one obtains $H_k(t) = (t - k)$ and

$$A = 1;\ B = 1;\ Q = 1;\ R = 1.3333$$

The discrete-time system is obviously controllable and observable, $R > 0$, and $Q - M'R^{-1}M = 0.8125 > 0$, so that the conditions for the existence of a steady-state solution to exist are met. With these discrete-time parameters substituted in the ARE (8.27), we obtain after some simplification the quadratic equation $0 = 1.0833 - \bar{P}^2$, with positive solution $\bar{P} = 1.0408$. With this value of $\bar{P}$ substituted in (8.28) we finally obtain $K = 0.6490$. The optimal "digital controller" is then given by $u_k^* = -0.6490x_k$, and the minimal value of V by $V^* = 1.0408x_0^2$, where $x_0 = x(0)$.

8.5 MATLAB Software

The MATLAB function **lqrd** can be used to compute the discrete-time feedback-gain matrix given by (8.28) for the continuous-time data $\tilde{A}, \tilde{B}, \tilde{Q}$, and $\tilde{R}$, and the sampling time T. In particular,

$$[K, \bar{P}, E] = lqrd(\tilde{A}, \tilde{B}, \tilde{Q}, \tilde{R}, T)$$

will return

- the optimal feedback matrix K such that $u_k = -Kx_k$
- the optimal performance matrix P such that $V^* = x_k' P x_k$
- the eigenvalues of the closed-loop system matrix $A - BK$ given by the entries of the vector E

To illustrate, we present the following example:

Example 8.2 Consider a typical process-control problem, a stirred tank with two input flow rates (and concentrations) and one output flow rate (and concentration). In particular, using the numerical values of example 1.2 of [113], we have the following linearized dynamics of the stirred tank problem

$$\begin{aligned} \dot{x} &= \begin{bmatrix} -0.01 & 0 \\ 0 & -0.02 \end{bmatrix} x + \begin{bmatrix} 1 & 1 \\ -0.25 & 0.75 \end{bmatrix} u; \\ y &= \begin{bmatrix} 0.01 & 0 \\ 0 & 1 \end{bmatrix} x \end{aligned}$$

where the incremental variables are defined as follows

- x_1 is the volume in the tank in cubic meters (cm)
- x_2 is the output concentration (kmol/cm)
- u_1 is the flow rate for input concentration 1 (cm/s)
- u_2 is the flow rate for input concentration 2 (cm/s)
- y_1 is the output flow rate (cm/s)
- y_2 is the output concentration (kmol/cm)

The problem is to design a digital-control system that minimizes the LQ performance measure

$$V = \int_0^\infty (y' \begin{bmatrix} 10^4 & 0 \\ 0 & 1 \end{bmatrix} y + u'u)dt$$

where the sampling time is $T = 10$ seconds. Note that the above performance measure corresponds to $\tilde{Q} = \tilde{R} = I_{2\times2}$. For the data of this problem, we obtain

$$K = \begin{bmatrix} 0.0897 & -0.1133 \\ 0.0318 & 0.1146 \end{bmatrix}$$

$$\bar{P} = \begin{bmatrix} 2.9035 & -0.0734 \\ -0.0734 & 3.0137 \end{bmatrix}$$

If required, the discrete-time matrices A and B may be computed from the MATLAB function **c2d**. In particular,

$$[A, B] = c2d(\tilde{A}, \tilde{B}, T)$$

returns the matrices A and B for a zero-order hold. For the above data, we find

$$A = \begin{bmatrix} 0.9048 & 0 \\ 0 & 0.8187 \end{bmatrix}$$

$$B = \begin{bmatrix} 9.5163 & 9.5163 \\ -2.2659 & 6.7976 \end{bmatrix}$$

△

For data given directly in the discrete-time domain, e.g.,

$$\begin{aligned} x_{k+1} &= Ax_k + Bu_k + \zeta_k \\ y_k &= Cx_k + \theta_k \\ V &= \lim_{k\to\infty} E\{x_k'Qx_k + u_k'Ru_k\} \end{aligned}$$

The steady-state LQG problem can be solved by the separation principle and the MATLAB functions **dlqr** and **dlqe**. In particular,

$$[K_c, S, E] = dlqr(A, B, Q, R)$$

returns the discrete-time, state-estimate feedback gain K_c such that $u_k = -K_c\hat{x}_k$ and

$$[K_f, S, E] = dlqe(A, B, C, \Psi, \Theta)$$

returns the Kalman-Bucy filter gain K_f such that

$$\hat{x}_{k+1} = A\hat{x}_k + Bu_k + K_f(y_k - C\hat{x}_k)$$

The solution of the discrete-time control and filter Riccati equations are given by S and P, respectively. The matrix E in the **dlqr** output is the closed-loop matrix $A - BK_c$, and the matrix E in the **dlqe** output is the matrix $A - K_fC$.

8.6 Notes and References

While linear-quadratic digital-control theory began to develop in the late fifties (see, e.g., Kalman and Bertram [102]), largely in response to new developments in digital computers, it was not until the early seventies, with the advent of microprocessors, that digital control assumed the importance it currently enjoys. Textbooks on the subject (e.g., Franklin, Powers, and Workman [70] or Lewis [121]), now include considerable details on microprocessor or digital-signal-processor implementation of digital controllers. See also Dorato and Petersen [54] and Moroney [146] for more on microprocessor implementation of digital controllers, and Hewer [88] for an algorithm to compute the steady-state feedback-gain matrix.

In this chapter we have focused on the theory for digital LQR control. This is essentially the theory first developed by Kalman and Koepke in 1958 [104]. Extensions to digital LQG theory may be found in the texts of Anderson and Moore [4], Kwakernaak and Sivan [113], and Lewis [121]. Stability margins for the discrete-time LQR solution are discussed in Shaked [184]. More details on the numerical solution and the theory of the discrete-time algebraic Riccati equations may be found in references Pappas, Laub, and Sandell [156] and Payne and Silverman [160], respectively.

In Mäkilä and Toivonen [130], a number of computational methods for the solution of the discrete-time LQG problem with a fixed-order compensator are surveyed, and in Mäkilä [128], algorithms are presented for LQG guaranteed-cost design for discrete-time systems when the plant uncertainty is modeled by a finite set of possible plants (multiple models).

For a comprehensive treatment of digital-control systems see the two volumes of Isermann [95], and for applications of digital control to robots and vehicles see Fargeon [62]. We have not explored LQ robust design for discrete-time systems in this brief chapter on digital control. An extensive discussion of discrete-time LQG/LTR theory may be found in chap. 5 of Bitmead, Gevers, and Wertz [29]. See also, Maciejowsky [126]. The discrete-time singular LQR problem is discussed in Peng and Kinnaert [161].

8.7 Problems

Problem 8.1 Find a state-feedback digital controller that minimizes

$$V = \int_0^\infty [(x_1)^2 + u^2]dt$$

for an analog plant with dynamics, $\dot{x}_1 = x_2, \dot{x}_2 = u$ and sampling interval $T = 1$.

Problem 8.2 Consider the same data as in problem 8.1, but with the upper limit in the performance measure replaced by the finite value $t_f = 5$. Use (8.18) and the discrete-time Riccati equation (8.19) to compute K_4 and K_3.

Problem 8.3 Show that when $M = 0$ and $R = 0$, the discrete-time Riccati equation (8.19) can be written

$$P_i = (A + BG_i)' P_{i+1} (A + BG_i) + Q$$

where

$$G_i = -(B' P_{i+1} B)^{-1} (B' P_{i+1} A)$$

Problem 8.4 Consider a discrete-time problem where $R = 0, M = 0$, and $S = Q > 0$. Show that if B is a matrix of full rank, then P_i exists and is positive definite. Hint: Use the results of problem 8.3 and mathematical induction, i.e., show that the result is true for P_N and that if $P_{i+1} > 0$, then $P_i > 0$.

Problem 8.5 Show that the optimal closed-loop, discrete-time system found in problem 8.4 is also asymptotically stable. Hint: Use the discrete-time Lyapunov function $V_i = x_i' P_i x_i$ to show that $V_{i+1} - V_i < 0$ for the closed-loop system $x_{i+1} = (A - BK_i)x_i$.

Problem 8.6 Design an optimal LQR digital state-feedback controller for the aircraft model in example 2.6. Assume $T = 0.1$ seconds, and use the matrices A, B, C given in example 2.6, with $Q = C'C$ and $R = I_{3\times 3}$. Plot the discrete-time signals for altitude, forward speed, and pitch angle, and compare with the analog signals of Figure 2.2.

Bibliography

[1] K. T. Alfriend (Ed.), Special section "Robust control design for a benchmark problem," *AIAA J. Guidance, Control and Dynamics* 15 (Sept.–Oct. 1992): 1060–1149.

[2] B.D.O. Anderson, N. K. Bose, and E. I. Jury, "Output feedback stabilization and related problems," *IEEE Trans. Automat. Contr.* AC-20 (Feb. 1975): 53–66.

[3] B.D.O. Anderson and J. B. Moore, *Linear Optimal Control.* Englewood Cliffs, NJ: Prentice Hall, 1971.

[4] B.D.O. Anderson and J. B. Moore, *Optimal Control: Linear Quadratic Methods.* Englewood Cliffs, NJ: Prentice Hall, 1990.

[5] B.D.O. Anderson and J. B. Moore, *Optimal Filtering.* Englewood Cliffs, NJ: Prentice Hall, 1979.

[6] B.D.O. Anderson and R. W. Scott, "Output feedback stabilization: Solution by algebraic geometry methods," *Proc. IEEE* 65 (1977): 849–861.

[7] L. Arnold, *Stochastic Differential Equations: Theory and Applications.* New York: Wiley, 1974.

[8] W. F. Arnold III and A. J. Laub, "Generalized eigenproblem algorithms and software for algebraic Riccati equations," *Proc. IEEE* 72 (1984): 1746–1754.

[9] M. Athans, "The matrix minimum principle," *Information and Control* 11 (1968): 592–606.

[10] M. Athans, "The role and use of the stochastic linear quadratic Gaussian problem in control system design," *IEEE Trans. Automat. Contr.* AC-16 (Dec. 1971): 529–552.

[11] M. Athans (Ed.), Special issue on linear-quadratic-gaussian problem, *IEEE Trans. Automat. Contr.* AC-16 (Dec. 1971).

[12] M. Athans and P. L. Falb, *Optimal Control.* New York: McGraw-Hill, 1966.

[13] M. Athans, P. Kapasouris, E. Kappos, and H. A. Spang, "Linear-quadratic-gaussian with loop transfer recovery methodology for the F-100 engine," *J. Guid., Contr. and Dynamics* 9 (Jan.-Feb. 1986): 45–52.

[14] J. G. Balchen and K. I. Mumme, *Process Control: Structures and Applications.* New York: Van Nostrand, 1988.

[15] B. R. Barmish, "Necessary and sufficient conditions for quadratic stabilizability of an uncertain system," *J. Optimiz. Theory Appl.* 46 (1985): 399–408.

[16] S. Barnett, *Matrices in Control Theory.* London: Van Nostrand Reinhold, 1971.

[17] T. Başar and P. Bernhard, "Differential Games and Applications," *Lecture Notes in Control and Information Sciences,* vol. 119. Berlin: Springer-Verlag, 1989.

[18] R. E. Bellman, "The theory of dynamic programming," *Proc. Nat. Acad. Sci. USA* 38 (1952): 716–719. Also in *Bull. Amer. Math. Society* 60 (1954): 503–516.

[19] R. E. Bellman and S. E. Dreyfus, *Applied Dynamic Programming.* Princeton, NJ: Princeton University Press, 1962.

[20] D. J. Bender and A. J. Laub, "The linear-quadratic optimal regulator for descriptor systems," *IEEE Trans. Automat. Contr.* AC-32 (Aug. 1987): 672–688.

[21] G. Bengtsson and S. Lindahl, "A design scheme for incomplete state or output feedback with applications to boiler and power system control," *Automatica* 10 (1974): 15–30.

[22] D. S. Bernstein, "Robust static and dynamic output-feedback stabilization: Deterministic and stochastic perspectives," *IEEE Trans. Automat. Contr.* AC-32 (Dec. 1987): 1076–1084.

[23] D. S. Bernstein and W. M. Haddad, "LQG Control with $\mathcal{H}^\infty$ performances bound: A Riccati equation approach," *IEEE Trans. Automat. Contr.* AC-34 (Mar. 1989): 293–305.

[24] D. S. Bernstein and W. M. Haddad, "The optimal projection equations with Petersen-Hollot bounds: Robust stability and performance via fixed-order dynamic compensation for systems with structured real-valued parameter uncertainty," *IEEE Trans. Automat. Contr.* AC-33 (June 1988): 578–582.

[25] D. S. Bernstein and W. M. Haddad, "Robust stability and performance via fixed-order dynamic compensation with guaranteed cost bounds," *Mathematics of Control Signal Systems* 3 (1990): 139–163.

[26] D. S. Bernstein and D. C. Hyland, "The optimal projection equations for reduced-order modeling, estimation and control of linear systems with multiplicative white noise," *J. Optimiz. Theory Appl.* 58 (1988): 387–409.

[27] J. P. Birdwell and A. J. Laub, "Balanced singular values for LQG/LTR design," *Int. J. Control* 45 (1987): 939–950.

[28] J. Bismut, "Linear-quadratic optimal stochastic control with random coefficients," *SIAM J. Contr. and Optim.* 14 (1976): 419–444.

[29] R. R. Bitmead, M. Gevers, and V. Wertz, *Adaptive Optimal Control.* Englewood Cliffs, NJ: Prentice Hall, 1990.

[30] S. Bittanti (Ed.), *The Riccati Equation in Control, Systems, and Signals.* Bologna: Pitagora Editrice, 1989.

[31] S. Boyd, V. Balakrishnan, C. Barratt, N. Khraishi, X. Li, D. Meyer, and S. Norman, "A new CAD method and associated architectures for linear controllers," *IEEE Trans. Automat. Contr.* AC-33 (Mar. 1988): 268–283.

[32] S. Boyd, V. Balakrishnan, E. Feron, and L. El Ghaoui, "Control system analysis and synthesis via linear matrix inequalities," *Proc. 1993 American Control Conf.* (May 1993): 2147–2154.

[33] S.P. Boyd and C. H. Barratt, *Linear Controller Design: Limits of Performance.* Englewood Cliffs, NJ: Prentice Hall, 1991.

[34] S. Boyd, L. El Ghaoui, E. Feron, and V. Balakrishnan, *Linear Matrix Inequalities in Systems and Control.* Philadelphia, PA: SIAM, 1994.

[35] A. E. Bryson, Jr. and Y. C. Ho, *Applied Optimal Control.* New York: Hemisphere, 1975.

[36] R. S. Bucy, "The Riccati equation and its bounds," *J. Comput. Syst. Sci.* 6 (1972): 343–353.

[37] P. E. Caines, *Linear Stochastic Systems.* New York: Wiley, 1988.

[38] F. M. Callier and J. L. Willems, "Criterion for the convergence of the solution of the Riccati differential equation," *IEEE Trans. Automat. Contr.* AC-26 (1981): 1232–1242.

[39] C. Caratheòdory, *Calculus of Variations and Partial Differential Equations of First Order.* San Fransisco: Holden-Day, 1967.

[40] J. L. Casti, "The linear-quadratic control problem: Some recent results and outstanding problems," *SIAM Review* 22 (Oct. 1980): 459–485.

[41] S.S.L. Chang and T.K.C. Peng, "Adaptive guaranteed cost control of systems with uncertain parameters," *IEEE Trans. Automat. Contr.* AC-17 (Aug. 1972): 474–483.

[42] B. M. Chen, A. Saberi, and P. Sannuti, "Necessary and sufficient conditions for a nonminimum phase plant to have a recoverable target loop: A stable compensator design for LTR," *Automatica* 28 (1992): 493–507.

[43] B. M. Chen, A. Saberi, and P. Sannuti, "A new stable compensator design for exact and approximate loop transfer recovery," *Automatica* 27 (Mar. 1991): 257–280.

[44] B.S. Chen and T.Y. Dong, "LQG optimal control design under plant perturbation and noise uncertainty: A state-space approach," *Automatica* 25 (1989): 431–436.

[45] C. T. Chen, *Linear System Theory and Design.* New York: Holt, Rinehart, and Winston, 1984.

[46] R. Y. Chiang and M. G. Safonov, *Robust-Control Toolbox for Use with* MATLAB. Natick, MA: MathWorks, 1988.

[47] D. J. Clements and B.D.O. Anderson, "Singular optimal control: The linear-quadratic problem," *Lecture Notes in Control and Information Sciences,* vol. 5. Berlin: Springer-Verlag, 1978.

[48] P. Colameri, R. Scattolini, and N. Schiavoni, "LQG optimal control of multirate sampled-data systems," *IEEE Trans. Automat. Contr.* AC-37 (May 1992): 675–682.

[49] E. J. Davison and A. Goldenberg, "The robust control of a general servomechanism problem: The servo compensator," *Automatica* 11 (1975): 461–471.

[50] P. Dorato, "Theoretical developments in discrete-time control," *Automatica* 19 (1983): 395–400.

[51] P. Dorato, L. Fortuna, and G. Muscato, "Robust control for unstructured perturbations: An introduction," *Lecture Notes in Control and Information Sciences,* vol. 168. Berlin: Springer-Verlag, 1992.

[52] P. Dorato and A. H. Levis, "Optimal linear regulators: the discrete-time case," *IEEE Trans. Automat. Contr.* AC-16 (Dec. 1971): 613–620.

[53] P. Dorato and Yunzhi Li, "U-parameter design of robust single-input-single-output systems," *IEEE Trans. Automat. Contr.* AC-36 (Sept. 1991): 971–975.

[54] P. Dorato and D. Petersen, "Digital control systems," in *Advances in Computers,* vol. 23, pp. 177–251. New York: Academic Press, 1984.

[55] P. Dorato and Tzu-Chung Yenn, "Robust LQG design via U-parameterization," in *Systems and Control,* pp. 39–49. Tokyo: Mita Press, 1991.

[56] J. C. Doyle, "Guaranteed margins for LQG regulators," *IEEE Trans. Automat. Contr.* AC-23 (Aug. 1978): 756–757.

[57] J. C. Doyle, B. A. Francis, and A. R. Tannenbaum, *Feedback Control Theory.* New York: Macmillan, 1992.

[58] J. C. Doyle, K. Glover, P. P. Khargonekar, and B. A. Francis, "State-space solutions to standard $\mathcal{H}_2$ and $\mathcal{H}_\infty$ control problems," *IEEE Trans. Automat. Contr.* AC-34 (Aug. 1989): 831–847.

[59] J. C. Doyle and G. Stein, "Multivariable feedback design: Concepts for a classical/modern synthesis," *IEEE Trans. Automat. Contr.* AC-26 (Feb. 1981): 4–16.

[60] J. C. Doyle and G. Stein, "Robustness with observers," *IEEE Trans. Automat. Contr.* AC-24 (Aug. 1979): 607–611.

[61] R. M. Dressler and D. Tabak, "Satellite tracking by combined optimal estimation and control techniques," *IEEE Trans. Automat. Contr.* AC-16 (Dec. 1971): 833–840.

[62] C. Fargeon, *The Digital Control of Systems: Applications to Vehicles and Robots.* London: North Oxford Academic, 1989.

[63] A. F. Fath, "Computational aspects of the linear optimal regulator problem," *IEEE Trans. Automat. Contr.* AC-14 (Oct. 1969): 547–550.

[64] J. J. Florentine, "Optimal control of continuous-time, Markov, stochastic systems," *J. Electronics Control* 10 (1961): 473–488.

[65] B. A. Francis, "A course in $\mathcal{H}_\infty$ Control Theory," *Lecture Notes in Control and Information Sciences,* vol. 88. Berlin: Springer-Verlag, 1987.

[66] B. A. Francis, "The optimal linear-quadratic time-invariant regulator with cheap control," *IEEE Trans. Automat. Contr.* AC-24 (Aug. 1979): 616–621.

[67] B. A. Francis and K. Glover, "Bounded peaking in the optimal linear regulator with cheap control," *IEEE Trans. Automat. Contr.* AC-23 (Aug. 1978): 608–617.

[68] B. A. Francis and W. M. Wonham, "The internal model principle of control theory," *Automatica* 12 (1976): 457–465.

[69] G. F. Franklin, J. D. Powell, and A. Emami-Naeni, *Feedback Control of Dynamic Systems.* New York: Addison-Wesley, 1986.

[70] G. F. Franklin, J. D. Powell, and M. L. Workman, *Digital Control of Dynamic Systems.* Reading, MA: Addison-Wesley, 1990.

[71] B. Friedland, *Control System Design.* New York: McGraw-Hill, 1986.

[72] P. Gahinet, A. Nemirovski, A.J. Laub,a nd M. Chilali, *LMI Toolbox: For use with* MATLAB, The MathWorks, Inc., Natick, MA, 1995.

[73] D. Gangsaas, "Application of modern synthesis to aircraft control: three case studies," *IEEE Trans. Automat. Contr.* AC-31 (Aug. 1986): 995–1104.

[74] F. R. Gantmacher, *Theory of Matrices* vol. 1. New York: Chelsea, 1959.

[75] M. Gevers and G. Li, *Parameterizations in Control, Estimation and Filtering Problems: Accuracy Aspects,* Communications and Control Engineering series. Berlin: Springer-Verlag, 1993.

[76] K. Glover, "All optimal Hankel-norm approximations of linear multivariable systems and their L^{∞}-error bounds," *Int. J. Control* 39 (1984): 1115–1193.

[77] K. Glover and J. C. Doyle, "State-space formulae for all stabilizing controllers that satisfy an H^{∞}-norm bound and relations to risk sensitivity," *Syst. Control Lett.* 11 (1988): 167–172.

[78] B. F. Goldstein, "Minimax control of linear unknown systems using mismatched state observers," *Int. J. Control* 20 (1974): 753–767.

[79] A. Grace, *Optimization Toolbox for Use with* MATLAB, Natick, MA: MathWorks, 1990.

[80] A. Grace, A. J. Laub, J. N. Little, and M. Thompson, *Control System Toolbox for Use with* MATLAB, Natick, MA: MathWorks, 1992.

[81] D. Yu. Grigor'ev and N. N. Vorobjov, "Solving systems of polynomial inequalities in subexponential time," *J. Symbolic Computation* 5 (1988): 37–64.

[82] M. Grimble and M. Johnson, *Optimal Control and Stochastic Estimation: Theory and Applications,* vols. 1 and 2. New York: Wiley, 1988.

[83] M. J. Grimble and T. J. Owens, "On improving the robustness of LQ regulators," *IEEE Trans. Automat. Contr.* AC-31 (Jan. 1986): 54–55.

[84] T. F. Gunckel and G. F. Franklin, "A general solution for linear sampled-data control systems," *J. Basic Eng. Trans. ASME* 85D (1963): 197–203.

[85] C. A. Harvey and G. Stein, "Quadratic weights for asymptotic regulator properties," *IEEE Trans. Automat. Contr.* AC-23 (June 1978): 378–387.

[86] U. Haussmann, "Optimal stationary control with state and control dependent noise," *SIAM J. Contr.* 9 (1971): 184–198.

[87] U. G. Haussmann, "Stability of linear systems with control dependent noise," *SIAM J. Contr.* 11 (1973): 382–394.

[88] G. A. Hewer, "An iterative technique for the computation of the steady state gain for the discrete optimal regulator," *IEEE Trans. Automat. Contr.* AC-16 (Aug. 1971): 382–384.

[89] O. Hijab, *Stabilization of Control Systems.* Applications of Mathematics Series, vol. 20. Berlin: Springer-Verlag, 1987.

[90] K. L. Hitz and B.D.O. Anderson, "Iterative method of computing the limiting solution of the matrix Riccati differential equation," *Proc. IEEE* 119 (1972): 1402–1406.

[91] Y. S. Hung and A.G.J. MacFarlane, "Multivariable feedback: A quasi-classical approach," *Lecture Notes in Control and Information Sciences,* vol. 40. Berlin: Springer-Verlag, 1982.

[92] D. C. Hyland and D. S. Bernstein, "The majorant Lyapunov equation: A nonnegative matrix equation for guaranteed robust stability and performance of large scale systems," *IEEE Trans. Automat. Contr.* AC-32 (Nov. 1987): 1005–1013.

[93] D. C. Hyland and D. S. Bernstein, "The optimal projection equations for fixed-order dynamic compensation," *IEEE Trans. Automat. Contr.* AC-29 (Nov. 1984): 1034–1037.

[94] F. Incertis, "A new formulation of the algebraic Riccati problem," *IEEE Trans. Automat. Contr.* AC-26 (June 1981): 768–770.

[95] R. Isermann, *Digital Control Systems,* vols. 1 and 2. Berlin: Springer-Verlag, 1989.

[96] D. H. Jacobson, *Extensions of Linear Quadratic Control.* New York: Academic Press, 1977.

[97] D. H. Jacobson, "Totally singular quadratic minimization problems," *IEEE Trans. Automat. Contr.* AC-16 (Dec. 1971): 651–658.

[98] S. M. Joshi and P. G. Maghami, "Robust compensators for flexible spacecraft control," *IEEE Trans. Aero. and Electr. Syst.* 28 (July 1992): 768–774.

[99] T. Kailath, *Linear Systems.* Englewood Cliffs, NJ: Prentice Hall, 1980.

[100] R. E. Kalman, "Contribution to the theory of optimal control," *Bol. Soc. Matem. Mex.* 5 (1960): 102-119.

[101] R. E. Kalman, "When is a linear control system optimal?," *Trans. ASME Ser. D (J. Basic Engr.)* 86 (1964): 51-60.

[102] R. E. Kalman and J. E Bertram, "General synthesis procedures for computer control of single and multiloop linear systems," *Trans. AIEE* 77, part 2 (1958): 602–609.

[103] R. E. Kalman and R. S. Bucy, "New results in linear filtering and prediction theory," *Trans. ASME Ser. D. (J. Basic Engr.)* 83 (1961): 95–107.

[104] R. E. Kalman and R. W. Koepcke, "Optimal synthesis of linear sampling control systems using generalized performance indices," *Trans. ASME Ser. D. (J. Basic Engr.)* 80 (1958): 1820–1826.

[105] J. P. Keller and B.D.O. Anderson, "A new approach to the discretization of continuous-time controllers," *IEEE Trans. Automat. Contr.* AC-37 (Feb. 1992): 214–223.

[106] P. P. Khargonekar and M. A. Rotea, "Mixed $\mathcal{H}^2/\mathcal{H}^\infty$ control: A convex optimization approach," *IEEE Trans. Automat. Contr.* AC-36 (July 1991): 824–837.

[107] P. P. Khargonekar and M. A. Rotea, "Multiple objective optimal control of linear systems: The quadratic norm case," *IEEE Trans. Automat. Contr.* AC-36 (Jan. 1991): 14–24.

[108] D. L. Kleinman, "On an iterative technique for Riccati equation computation," *IEEE Trans. Automat. Contr.* AC-13 (Feb. 1968): 114–115.

[109] D. L. Kleinman, "Optimal stationary control of linear systems with control-dependent noise," *IEEE Trans. Automat. Contr.* AC-14 (Dec. 1969): 673–677.

[110] V. Kučera, *Discrete Linear Control.* New York: Wiley, 1979.

[111] H. J. Kushner, *Introduction to Stochastic Control.* New York: Holt, Rinehart, and Winston, 1971.

[112] H. J. Kushner, *Stochastic Stability and Control.* New York: Academic Press, 1967.

[113] H. Kwakernaak and R. Sivan, *Linear Optimal Control Systems.* New York: Wiley Interscience, 1972.

[114] P. Lancaster and L. Rodman, "Existence and uniqueness theorems for the algebraic Riccati equation," *Int. J. Control* 32 (1980): 285–309.

[115] A. J. Laub, "A Schur method for solving algebraic Riccati equations," *IEEE Trans. Automat. Contr.* AC-24 (Dec. 1979): 913–921.

[116] B. K. Lee, B. S. Chen, and Y. P. Lin, "Extension of linear quadratic optimal control theory for mixed backgrounds," *Int. J. Control* 54 (1991): 943–972.

[117] T. T. Lee and M. S. Chen, "Robustness recovery of LQG-based multivariable control designs," *Int. J. Control* 45 (1987): 1131–1136.

[118] N. A. Lehtomaki, N. R. Sandell, Jr., and M. Athans, "Robustness results in linear quadratic Gaussian based multivariable control," *IEEE Trans. Automat. Contr.* AC-26 (Feb. 1981): 75–92.

[119] A. M. Letov, "Analytic controller design I, II," *Autom. Remote Contr.* 21 (1960): 303–306.

[120] W. S. Levine, T. L. Johnson, and M. Athans, "Optimal limited state variable feedback controllers for linear systems," *IEEE Trans. Automat. Contr.* AC-16 (Dec. 1971): 785–793.

[121] F. L. Lewis, *Applied Optimal Control and Estimation.* Englewood Cliffs, NJ: Prentice Hall, 1992.

[122] F. L. Lewis, *Optimal Control.* New York: Wiley, 1986.

[123] F. L. Lewis, *Optimal Estimation: With an Introduction to Stochastic Control Theory*, New York: Wiley, 1986.

[124] D. G. Luenberger, *Linear and Nonlinear Programming.* New York: Addison-Wesley, 1984.

[125] A. MacFarlane, "An eigenvector solution of the optimal linear regulator problem," *J. Electr. and Control* 14 (1965): 643–654.

[126] J. M. Maciejowski, "Asymptotic recovery for discrete-time systems," *IEEE Trans. Automat. Contr.* AC-30 (June 1985): 602–605.

[127] A. N. Madiwale, W. M. Haddad, and D. S. Bernstein, "Robust H^∞ control design for systems with structured parameter uncertainty," *Syst. Control Lett.* 12 (1988): 393–407.

[128] P. M. Mäkilä, "Multiple models, multiplicative noise and linear quadratic control-algorithm aspects," *Int. J. Control* 54 (1991): 921–941.

[129] P. M. Mäkilä, "On multiple criteria stationary linear quadratic control," *IEEE Trans. Automat. Contr.* 34 (Dec. 1989): 1311–1313.

[130] P. M. Mäkilä and H. T. Toivonen, "Computational methods for parametric LQ problems: A survey," *IEEE Trans. Automat. Contr.* 32 (Aug. 1987): 658–671.

[131] K. Martensson, "On the matrix Riccati equation," *Information Sci.* 3 (1971): 17–49.

[132] MATLAB User's Guide. Natick, MA: MathWorks, 1989.

[133] C. L. Matson and P. S. Maybeck, "On an assumed convergence result in the LQG/LTR technique," *IEEE Trans. Automat. Contr.* AC-36 (Jan. 1991): 123–125.

[134] P. S. Maybeck, *Stochastic Models, Estimation, and Control,* vols. 1–3. New York: Academic Press, 1979.

[135] P. J. McLane, "Optimal stochastic control of linear systems with state and control-dependent disturbances," *IEEE Trans. Automat. Contr.* AC-16 (Dec. 1971): 793–798.

[136] M. D. McLaren and G. L. Slater, "Robust multivariable control of large space structures using positivity," *J. Guid., Contr. and Dynamics* 10 (1987): 393–400.

[137] D. McLean, *Automatic Flight Control Systems.* International Series in Systems and Control Engineering. London: Prentice Hall, 1990.

[138] J. Medanic, H. S. Tharp, and W. R. Perkins, "Pole placement by performance criterion modification," *IEEE Trans. Automat. Contr.* AC-33 (May 1988): 469–472.

[139] V. L. Mehrmann, "The autonomous linear quadratic control problem," *Lecture Notes in Control and Information Sciences,* vol. 163. Berlin: Springer-Verlag, 1991.

[140] J. M. Mendel and D. L. Geseking, "Bibliography on the linear-quadratic-gaussian problem," *IEEE Trans. Automat. Contr.* AC-16 (Dec. 1971): 847–869.

[141] G. Menga, M. Milanese, and A. L. Negro, "Min-max quadratic cost control of systems described by approximate models," *IEEE Trans. Automat. Contr.* AC-21 (Oct. 1976): 651–659.

[142] M. Mercadal, "Homotopy approach to optimal, linear quadratic, fixed architecture compensation," *J. Guid., Contr. and Dynamics* 14 (1991): 1224–1233.

[143] H. B. Meyer, "The matrix equation $AZ + B - ZCZ - ZD = 0$," *SIAM J. Appl. Math.* 30 (1976): 136–142.

[144] B. P. Molinari, "The time invariant linear quadratic optimal control problem," *Automatica* 13 (1977): 347–457.

[145] J. B. Moore and L. Xia, "Loop recovery and robust state estimate feedback design," *IEEE Trans. Automat. Contr.* AC-36 (June 1987): 512–517.

[146] P. Moroney, *Issues in the Implementation of Digital Feedback Compensators.* Cambridge, MA: MIT Press, 1983.

[147] D. Mustafa, "Combined $\mathcal{H}^\infty$/LQG control via the optimal projection equations: On minimizing the LQG cost bound," *Int. J. of Robust and Nonlinear Control* 1 (1991): 99–110.

[148] D. Mustafa, "Relations between maximum-entropy/$\mathcal{H}^\infty$ control and combined $\mathcal{H}^\infty$/LQG control," *Syst. Control Lett.* 12 (1989) 193–203.

[149] D. Mustafa and K. Glover, "Minimum entropy $\mathcal{H}^\infty$ control," *Lecture Notes in Control and Information Sciences,* vol. 146. Springer-Verlag, 1990.

[150] Yu. Nesterov and A. Nemirovsky, *Interior Point Polynomial Methods in Convex Programming: Theory and Applications.* Philadelphia, PA: SIAM, 1993.

[151] G. C. Newton, Jr., L. A. Gould, and J. F. Kaiser *Analytic Design of Linear Feedback Controls.* New York: Wiley, 1957.

[152] K. Nordström, "Trade-off between noise sensitivity and robustness for LQG regulators," *Int. J. Control* 46 (1987): 1689–1714.

[153] A. V. Oppenheim and R. W. Schafer, *Digital Signal Processing.* Englewood Cliffs, NJ: Prentice Hall, 1975.

[154] A. V. Oppenheim and R. W. Schafer, *Discrete-Time Signal Processing.* Englewood Cliffs, NJ: Prentice Hall, 1989.

[155] B. T. Oranç and C. L. Phillips, "On the accuracy of the stochastic simulation of infinite-horizon LQG control systems," *IEEE Trans. Automat. Contr.* AC-36 (Apr. 1991): 492–494.

[156] T. Pappas, A. J. Laub, and N. R. Sandell, "On the numerical solution of the discrete-time algebraic Riccati equation," *IEEE Trans. Automat. Contr.* AC-25 (Aug. 1980): 631–641.

[157] K. T. Parker, "Design of PID controllers by use of optimal linear regulator theory," *Proc. IEE, E* 119 (1972): 911–914.

[158] R. V. Patel and N. Munro, *Multivariable System Theory and Design.* Oxford: Pergamon Press, 1982.

[159] R. V. Patel, M. Toda, and S. Sridhar, "Robustness of linear quadratic state feedback designs in the presence of system uncertainty," *IEEE Trans. Automat. Contr.* AC-22 (Dec. 1977): 945–949.

[160] H. J. Payne and L. M. Silverman, "On the discrete time algebraic Riccati equation," *IEEE Trans. Automat. Contr.* AC-18 (June 1973): 226–234.

[161] Y. Peng and M. Kinnaert, "Explicit solution to the singular LQ regulation problem," *IEEE Trans. Automat. Contr.* AC-37 (May 1992): 633–636.

[162] I. R. Petersen, "Complete results for a class of state feedback disturbances attenuation problems," *IEEE Trans. Automat. Contr.* AC-34 (Nov. 1989): 1196–1199.

[163] I. R. Petersen, "A Riccati equation approach to the design of stabilizing controllers and observers for a class of uncertain systems," *IEEE Trans. Automat. Contr.* AC-30 (Sept. 1985): 904–907.

[164] I. R. Petersen, "A stabilization algorithm for a class of uncertain linear systems," *Syst. Control Lett.* 8 (1988): 351–357.

[165] I. R. Petersen, "Stabilization of an uncertain linear system in which uncertain parameters enter into the input matrix," *SIAM J. Contr. and Optim.* 26 (1988): 1257–1264.

[166] I. R. Petersen and C. V. Hollot, "A Riccati equation approach to the stabilization of uncertain linear systems," *Automatica* 22 (1986): 397–411.

[167] P. H. Petkov, N. D. Christov, and M. M. Konstantinov, "On the numerical properties of the Schur approach for solving the matrix Riccati equation," *Syst. Control Lett.* 9 (1987): 197–201.

[168] Y. A. Phillis, "Controller design of systems with multiplicative noise," *IEEE Trans. Automat. Contr.* AC-30 (Oct. 1985): 1017–1019.

[169] A.C.M. Ran and L. Rodman, "The algebraic matrix Riccati equation," *Operator Theory: Advances and Applications* 12 (1984): 351–381.

[170] L. R. Ray, "Stability robustness of uncertain LQG/LTR systems," *IEEE Trans. Automat. Contr.* AC-38 (Feb. 1993): 304–308.

[171] W. Harmon Ray, *Advanced Process Control.* New York: McGraw Hill, 1981.

[172] W. T. Reid, *Riccati Differential Equations.* New York: Academic Press, 1972.

[173] D. B. Ridgely, S. S. Banda, T. E. McQuade, and P. J. Lynch, "An application of linear-quadratic-Gaussian with loop transfer recovery to an unmanned aircraft," *J. Guid., Contr. and Dynamics* 10 (1987): 82–90.

[174] M. A. Rotea and P. P. Khargonekar, "$\mathcal{H}^2$-optimal control with an $\mathcal{H}^\infty$-constraint: The state feedback case," *Automatica* 27 (1991): 307–316.

[175] A. Saberi, B. M. Chen, and P. Sannuti, *Loop Transfer Recovery: Analysis and Design.*' Communications and Control Engineering series. Berlin: Springer-Verlag, 1993.

[176] A. Saberi, B. M. Chen, P. Sannuti, and U. L. Ly, "Simultaneous H_2/H_∞ optimal control: The state feedback case," *Automatica* 29 (1993): 1611–1614.

[177] A. Saberi and P. Sannuti, "Cheap and singular control for linear quadratic regulators," *IEEE Trans. Automat. Contr.* AC-32 (Mar. 1987): 208–219.

[178] M. Saeki, "$\mathcal{H}^\infty$/LTR procedure with specified degree of recovery," *Automatica* 28 (1992): 509–517.

[179] M. G. Safonov, *Stability and Robustness of Multivariable Feedback Systems.* Cambridge, MA: MIT Press, 1980.

[180] M. G. Safonov and M. Athans, "Gain and phase margins of multiloop LQG regulators," *IEEE Trans. Automat. Contr.* AC-22 (Apr. 1977): 173–179.

[181] M. G. Safonov and M. Athans, "A multiloop generalization of the circle criterion for stability margin analysis," *IEEE Trans. Automat. Contr.* AC-26 (Apr. 1981): 415–421.

[182] M. G. Safonov, A. J. Laub, and G. L. Hartmann, "Feedback properties of multivariable systems: The role and use of the return difference matrix," *IEEE Trans. Automat. Contr.* AC-26 (Feb. 1981): 47–65.

[183] B. Shafai, S. Beale, A. H. Nieman, and J. Stoustrup, "Modified structures for loop transfer recovery design," *Proc. 1993 Conference on Decision and Control* (1993): 3341-3344.

[184] U. Shaked, "Guaranteed stability margins for the discrete-time linear quadratic optimal regulator," *IEEE Trans. Automat. Contr.* AC-31 (Feb. 1986): 162–165.

[185] U. Shaked and E. Soroka, "On the stability robustness of the continuous-time LQG optimal control," *IEEE Trans. Automat. Contr.* AC-30 (Oct. 1985): 1039–1043.

[186] H. A. Shubert, "Analytic solution for the algebraic Riccati equation," *IEEE Trans. Automat. Contr.* AC-19 (June 1974): 255–256.

[187] E. Soroka and U. Shaked, "On the robustness of LQ regulators," *IEEE Trans. Automat. Contr.* AC-29 (July 1984): 664–665.

[188] C. E. de Souza and M. D. Fragoso, "On the existence of maximal solution for generalized algebraic Riccati equations arising in stochastic control," *Syst. Control Lett.* 14 (1990): 233–239.

[189] G. Stein, "Generalized quadratic weights for asymptotic regulator properties," *IEEE Trans. Automat. Contr.* AC-24 (Aug. 1979): 559–566.

[190] G. Stein and M. Athans, "The LQG/LTR procedure for multivariable feedback control design," *IEEE Trans. Automat. Contr.* AC-32 (Feb. 1987): 105–114.

[191] R. F. Stengel, *Stochastic Optimal Control.* New York: Wiley, 1986.

[192] H. T. Toivonen, "A multiobjective linear quadratic Gaussian control problem," *IEEE Trans. Automat. Contr.* AC-29 (Mar. 1984): 279–280.

[193] M. Tomizuka and D. Janczak, "Linear quadratic design of decoupled preview controllers for robotic arms," *Int. Jour. Robotics Research* 4 (Spring 1985): 67–74.

[194] A. Trofino-Neto, J. M. Dion, and L. Dugard, "Robustness bounds for LQ regulators," *IEEE Trans. Automat. Contr.* AC-37 (Sept. 1992): 1373–1377.

[195] A. Trofino-Neto and V. Kučera, "Stabilization via static output feedback," *IEEE Trans. Automat. Contr.* AC-38 (May 1993): 764–765.

[196] S. C. Tsay, I. K. Fong, and T. S. Kuo, "Robust linear quadratic optimal control for systems with linear uncertainties," *Int. J. Control* 53 (1991): 81–96.

[197] J. S. Tyler and F. B. Tuteur, "The use of a quadratic performance index to design multivariable control systems," *IEEE Trans. Automat. Contr.* AC-11 (Jan. 1966): 84–92.

[198] A. G. Ulsoy and D. Hrovat, "Stability robustness of LQG active suspensions," *Proc. 1990 American Control Conf.* (May 1990): 1347–1356.

[199] M. Vidyasagar, *Control System Synthesis: A Factorization Approach.* Cambridge: MA, MIT Press, 1985.

[200] T. L. Vincent and W. J. Grantham, *Optimality in Parametric Systems.* New York: Wiley, 1981.

[201] A. Vinkler and L. J. Wood, "Multistep guaranteed cost control of linear systems with uncertain parameters," *AIAA J. Guidance, Control and Dynamics* 2 (1979): 449–456.

[202] B. Wie and D. S. Bernstein, "A benchmark problem for robust control design," *Proc. 1993 American Control Conf.* (May 1990): 961–962.

[203] N. Wiener, *Cybernetics.* New York: Wiley, 1948.

[204] N. Wiener, *Extrapolation, Interpolation, and Smoothing of Stationary Time Series.* Cambridge, MA: Technology Press, 1949.

[205] J. C. Willems, "Least squares stationary optimal control and the algebraic Riccati equation," *IEEE Trans. Automat. Contr.* AC-16 (Dec. 1971): 621–634.

[206] J. C. Willems, "On the existence of a nonpositive solution to the Riccati equation," *IEEE Trans. Automat. Contr.* AC-19 (Oct. 1974): 592–593.

[207] J. L. Willems and J. C. Willems, "Feedback stabilizability for stochastic systems with state and control dependent noise," *Automatica* 12 (1976): 277–283.

[208] E. Wong and M. Zakai, "On the relation between ordinary and stochastic differential equations," *Int. J. Engr. Sci.* 3 (1965): 213–229.

[209] W. M. Wonham, "On a matrix equation of stochastic control," *SIAM J. Contr.* 6 (1968): 681–697.

[210] W. M. Wonham, "On the separation theorem of stochastic control," *SIAM J. Contr.* 6 (1968): 312–326.

[211] W. M. Wonham, "Optimal stationary control of a linear system with state-dependent noise," *SIAM J. Contr.* 5 (1967): 486–500.

[212] W. M. Wonham, "Random differential equations in control," *Probabilistic Methods in Applied Mathematics,* vol. 2. A.T. Bharucha-Reid (Ed.) New York: Academic Press, 1970.

[213] D. C. Youla, J. J. Bongiorno, Jr., and H. A. Jabr, "Modern Wiener-Hopf design of optimal controllers: Parts I, II," *IEEE Trans. Automat. Contr.* AC-21 (1976): 3–14; and AC-21 (1976): 319–330.

[214] Z. Zhang and J. S. Freudenberg, "Loop transfer recovery for nonminimum phase plant," *IEEE Trans. Automat. Contr.* AC-35 (May 1990): 547–553.

[215] K. Zhou and P. P. Khargonekar, "Robust stabilization of linear systems with norm-bounded time-varying uncertainty," *Syst. Control Lett.* 10 (1988): 17–20.

Index